River Basin Management X

WITPRESS

WIT Press publishes leading books in Science and Technology.
Visit our website for new and current list of titles.
www.witpress.com

WITeLibrary

Home of the Transactions of the Wessex Institute.
Papers published in this volume are archived in the WIT eLibrary in volume 234 of WIT Transactions on Ecology and the Environment (ISSN 1743-3541).
The WIT eLibrary provides the international scientific community with immediate and permanent access to individual papers presented at WIT conferences.
http://library.witpress.com

TENTH INTERNATIONAL CONFERENCE ON
RIVER BASIN MANAGEMENT

RIVER BASIN MANAGEMENT X

CONFERENCE CHAIRMEN

S. Mambretti
Polytechnic of Milan, Italy
Member of WIT Board of Directors

J. Melgarejo
University of Alicante, Spain

INTERNATIONAL SCIENTIFIC ADVISORY COMMITTEE

ORGANISED BY
Wessex Institute, UK
University of Alicante, Spain

SPONSORED BY
WIT Transactions on Ecology and the Environment
International Journal of Sustainable Development and Planning

WIT Transactions

Wessex Institute
Ashurst Lodge, Ashurst
Southampton SO40 7AA, UK

Senior Editors

Editorial Board

River Basin Management X

EDITORS

S. Mambretti
Polytechnic of Milan, Italy
Member of WIT Board of Directors

J. Melgarejo
University of Alicante, Spain

WIT*PRESS* Southampton, Boston

Editors:
S. Mambretti
Polytechnic of Milan, Italy
Member of WIT Board of Directors

J. Melgarejo
University of Alicante, Spain

Published by

WIT Press
Ashurst Lodge, Ashurst, Southampton, SO40 7AA, UK
Tel: 44 (0) 238 029 3223; Fax: 44 (0) 238 029 2853
E-Mail: witpress@witpress.com
http://www.witpress.com

For USA, Canada and Mexico

Computational Mechanics International Inc
25 Bridge Street, Billerica, MA 01821, USA
Tel: 978 667 5841; Fax: 978 667 7582
E-Mail: infousa@witpress.com
http://www.witpress.com

British Library Cataloguing-in-Publication Data

A Catalogue record for this book is available
from the British Library

ISBN: 978-1-78466-327-8
eISBN: 978-1-78466-328-5
ISSN: 1746-448X (print)
ISSN: 1743-3541 (on-line)

The texts of the papers in this volume were set individually by the authors or under their supervision. Only minor corrections to the text may have been carried out by the publisher.

Preface

Papers presented at the 10th in a series of conferences on River Basin Management are contained in this book. The included works mark a growing global interest in the planning, design and management of river basin systems and take in to account all aspects of Hydrology, Ecology, Environmental Management, Flood Plains and Wetlands.

Catastrophic events such as floods and associated landslides, erosion and sedimentation can have serious effects not only on life and property but also on the basin ecology. Frequently these problems are aggravated by the unforeseen consequences of man made changes in the river basin. This has led in recent years to work on river restoration and rehabilitation with various degrees of success.

Changes in the landscape, use of the land and climate conditions leads to a continuous revaluation of river basin management objectives requiring the development of better measuring tools in conjunction with accurate computer technology.

Specific themes covered in this volume include: Water resources management; Flood risk management; Ecological and environmental impact; Erosion and sediment transport; Hydrological modelling; River restoration and rehabilitation; Hydropower issues and development; River and watershed management; Water quality issues; Organic contamination management; Agricultural pollution; Transboundary water issues; Estuaries and deltas; Climate change; Remote sensing; Hydraulic structures; Rain water management; Water energy nexus; Drought assessment and management; Ecosystem services.

Papers presented at the River Basin Management meetings are permanently available to the international community in digital format at http://www.witpress.com/elibrary .

The Editors
2019

Contents

SEDIMENT TRANSPORT AT RIVER CONTRACTIONS

GIUSEPPE OLIVETO
School of Engineering, University of Basilicata, Italy

ABSTRACT
River contractions might occur because of either natural controls or hydraulic works. This paper aims to provide new insights on the temporal evolution of the morphological patterns at river contractions based on experiments at laboratory scale. Laboratory tests were performed in a 20 m long straight channel having a rectangular cross section, 1.0 m wide and 1.0 m deep. The erodible bed reach was made by almost uniform sediment consisting of either rough sand or medium gravel. The contraction model was 0.5, 1.0, 2.0, or 3.0 m long. Three contraction ratios $\beta = b/B$ = 0.9, 0.8, and 0.7 were considered, with b width of the contracted reach and B width of the approach channel. The experiments were performed in clear-water scour regime and steady flow conditions. Typically, they were of long duration (up to 15 days) to achieve conditions of dynamic equilibrium, but runs of short duration were also carried out to acquire suitable experimental data at the earlier stages of the bed forms development. Compared to previous studies by the author on the same topic, new results are provided on bed morphology characteristics, also considering literature data, and predictive models on the time-dependent development of the bed degradation in and downstream of the contracted area.
Keywords: river contraction, scouring, sediment transport.

1 INTRODUCTION

In a paper of 1934, Straub [1] emphasized how river contractions typically occur when structures are built for narrowing the river so as to obtain a greater navigable bed. Gill [2] observes that the river channel within the parallel guide banks, built on diversion works such as weirs and barrages, is a practical example of rectangular long contraction. By way of example, he shows that at the site of the Mailsi Siphon, Pakistan, the natural width of the Sutlej river is reduced from 12 km to merely 432.0 m. In more general terms, Dey and Raikar [3] highlight that contractions of river width in building bridges, barrages, weirs, and cross-drainage works are common examples of channel contractions. Moreover, partial closure required for the maintenance of channel riverbanks, such as construction of cofferdam and end-dump channel constriction, is the other type of channel contraction.

The flow velocity in a channel contraction increases due to the reduction of the flow area. The bed shear stresses therefore increase, and hence scour processes may occur. Several predictive models have been proposed for estimating the maximum scour depth at channel contractions. A large part of these models are based on the DuBoys' transport formula and Manning's equation (e.g. Gill [2], Laursen [4], Komura [5], Lim [6], Lim and Cheng [7]). In 2005 Dey and Raikar [3] published a paper on long contractions with uniform and non-uniform sediments by making available the experimental data for 131 runs under clear-water scour. Later, Dey and Raikar [8] proposed an analytical model for the computation of live-bed scour depth in long contractions, based on the energy and continuity equations. The scour depths computed from the model were found in satisfactory agreement with experimental data. Previously proposed equations for estimation of the equilibrium scour depth were compared with experimental data and the comparison revealed that the equation proposed by Lim [6] was the best predictor among those compared. Brandimarte et al. [9] developed a probabilistic framework to calculate the probability density function and exceedance probability of the estimated contraction scour depth over the life of a bridge.

Compared to laboratory experimental studies, field investigations are still limited. As examples, Benedict [10] measured clear-water contraction scour in the South Carolina Piedmont; Mueller and Wagner [11] conducted an analysis of the available field data for contraction scour. They compared field data with estimates from the formulas of Straub, Laursen, and Komura. The results revealed overprediction in most cases, but instances of underprediction also occurred. Detailed real-time measurements of flow velocities and bed elevations for a flood in 1997 on the Pomme de Terre River in Minnesota are also provided. Hong and Abid [12] carried out laboratory experiments using a 1:45 scale hydraulic model of the Ocmulgee River Bridge at Macon, Georgia including upstream and downstream river reaches and bridge structure. The experimental studies elucidated the interaction between local pier scour and contraction scour. Comparisons are made between scour measurements in the laboratory and in the field.

Nevertheless, current literature data are still insufficient to assess reliable equations especially when the effect of the time is included. More important, most of the literature papers consider conditions of dynamic equilibrium, and hence concentrate on the maximum scour depth, overlooking the analysis of the channel bed morphology in its entirety, both in and outside the contracted zone. Based on experiments at laboratory scale and compared to previous papers by the author on the same topic (Oliveto and Marino [13]–[15]), new results are given in terms of empirical equations on the peculiar bed morphology characteristics also considering literature data, and predictive models on the temporal development of the bed degradation in and downstream of the contracted area.

2 EXPERIMENTS

Experiments were made at the Hydraulic Engineering Laboratory, University of Basilicata, Italy, in a 1 m wide and 20 m long rectangular straight channel. In total 22 runs were performed.

The mobile bed was from 10 up to 16 m long depending on the contraction model length. Two bed sediments were tested either a nearly uniform (in size) sand with median grain size d_{50} = 1.7 mm, sediment gradation $\sigma = (d_{84}/d_{16})$ = 1.5, and dimensionless grain size $D^* = (g'/\nu^2)^{1/3} d_{50} = 43$ or a nearly uniform (in size) gravel with d_{50} = 9.0 mm, σ = 1.5, and D^* = 228. ν is the kinematic viscosity of water and g' is the modified gravitational acceleration equal to $g(\rho_s-\rho)/\rho$, with ρ mass density of water, ρ_s mass density of sediment, and g gravitational acceleration. The sand bed was used to simulate transition conditions at the interface sediment-water while the gravel bed hydraulically rough conditions, according to the Shields' diagram. The approaching mobile bed upstream of the contraction model was at least 5 m long to facilitate the development of fully turbulent flow. Whereas, the mobile bed downstream of the contraction model was at least 3 m long for allowing an undisturbed evolution of the bed morphology.

The contraction model was formed by means of three parts. The middle one was straight-lined of length l either 0.5, 1.0, 2.0, or 3.0 m; the upstream inlet transition and the downstream outlet transition were identical in shape and nearly streamlined to reduce energy losses and local scour effects. Three contraction ratios $\beta = b/B$ were tested either 0.9, 0.8, or 0.7, with b the width of the contraction model in its middle part and B the width of the uncontracted channel. Overall, the ratio of b, and hence of B, to the approach flow depth h_o was kept > 4 to minimize sidewall effects [16].

Further details on the experimental work, main test conditions and results for each run included, are provided in Oliveto and Marino [13]–[15]. With reference to the run 21 characterised by sandy movable bed, contraction ratio β = 0.70, contraction length

$l = 3.00$ m, discharge $Q = 0.040$ m^3/s, and approach flow depth $h_o = 0.100$ m, Fig. 1 shows the free surface features and the bed morphology at the end of the run just after the channel was drained.

(a) (b)

Figure 1: Run 21 with contraction ratio $\beta = 0.70$, contraction length $l = 3.00$ m, discharge $Q = 0.040$ m^3/s, and approach flow depth $h_o = 0.100$ m. (a) Free surface; and (b) Bed morphology at the end of the run.

3 ANALYSIS OF DATA AND RESULTS

In sediment transport processes at channel contractions, the maximum scour depth z^* at a given time t can be expressed in functional form as:

$$z^* = f(V_o, h_o, \nu, \rho, \rho_s, d_{50}, \sigma, B, b, l, g, t), \tag{1}$$

in which V_o is the approach flow velocity, h_o is the approach flow depth, ν is the kinematic viscosity, ρ is the mass density of the water, ρ_s is the mass density of the sediment, and g is the gravitational acceleration.

In two-phase flow phenomena with sediment–water interaction, it is appropriate to group the independent variables g, ρ, and ρ_s into $g' = g(\rho_s - \rho)/\rho$ as done by [3], [17]. Also, the contraction ratio $\beta = b/B$ can be introduced in order to account for the combined effect of b and B. In addition, the influence of ν in turbulent flow over a rough bed can be considered negligible [3]. Hence, the functional eqn (1) simplifies as

$$z^* = f(V_o, h_o, d_{50}, \sigma, \beta, l, g', t). \tag{2}$$

Applying the Buckingham π theorem with repeating variables h_o and V_o, and rearranging the dimensionless parameters in a similar way as in [17], yields

$$Z^* = \frac{z^*}{h_o} = \left(F_d, \frac{h_o}{d_{50}}, \sigma, \beta, \frac{l}{h_o}, T\right), \tag{3}$$

with F_d the densimetric Froude number defined as $V_o/(g'd_{50})^{1/2}$ and T a dimensionless time defined as $(g'd_{50})^{1/2}t/h_o$. The same functional relationship applies to $L^* = l^*/h_o$, in which l^* is the scour hole length.

The experimental data collected at the University of Basilicata and those by Dey and Raikar [3] were used for regression analysis, which yields the following equation in the case of z^*

$$Z^* = 0.025\left(\frac{h_o}{d_{50}}\right)^{-0.16}\left(1+\frac{l}{h_o}\right)^{0.64}\left(\frac{1-\beta}{\beta}\right)^{1.02} F_d^{\,1.65}T^{0.09}, \tag{4}$$

with coefficient of determination $r^2 = 0.84$ between observed and computed values of z^* (Fig. 2(a)). Eqn (4) is consistent with the condition that Z^* is equal to zero for $\beta = 1$ and tends to infinity when β goes to zero. It is also consistent with the condition that Z^* should not necessarily be zero when $l = 0$. The sediment gradation σ was not considered in this analysis because all the bed sediments used in the experiments were nearly uniform.

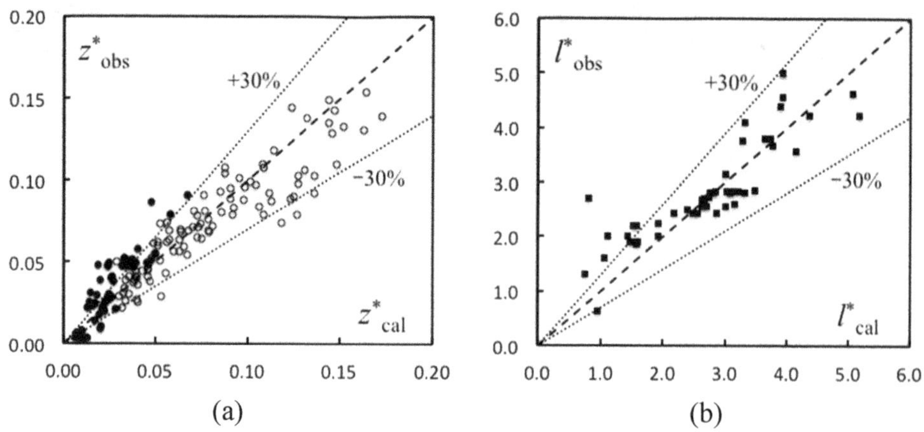

Figure 2: (a) Comparison between observed (subscript "obs") and calculated (subscript "cal") values of z^* for the data collected at the University of Basilicata (black symbols) and the data by Dey and Raikar [3] (white symbols). The values of z^* are in meters; (b) Comparison between observed and calculated values of l^* for the data collected at the University of Basilicata. The values of l^* are in meters.

Fig. 2(a) shows that the data points are reasonably well close to the line of perfect agreement with deviations generally within ±30%. However, the data collected at the University of Basilicata appear somewhat underestimated. This discrepancy between the two datasets may be explained by the fact that in the case of the data by Dey and Raikar [3] the exact duration of each run is unknown and here it was assumed always equal to 24 hours according to the authors' statement that the experiments were run for a period more than 24 hours. Moreover, the experiments by Dey and Raikar [3] were carried out in a somewhat narrow channel (0.6 m wide) which could lead to some scale effects. A slightly more conservative equation for Z^* could therefore be

$$Z^* = 0.025\left(\frac{h_o}{d_{50}}\right)^{-0.16}\left(1+\frac{l}{h_o}\right)^{0.64}\left(\frac{1-\beta}{\beta}\right)^{1.02} F_d^{\,1.65}T^{0.12}. \tag{5}$$

There are not systematic literature data in the case of l^*, then the experimental data collected at University of Basilicata only were considered for regression analysis, which yields the following equation

$$L^* = 1.244\left(\frac{h_o}{d_{50}}\right)^{-0.40}\left(1+\frac{l}{h_o}\right)^{0.80}\left(\frac{1-\beta}{\beta}\right)^{-0.19} F_d^{\,1.87} T^{0.08}, \tag{6}$$

with coefficient of determination $r^2 = 0.69$ between observed and calculated values of l^* (Fig. 2(b)). Interestingly, L^* was found practically independent of the contraction ratio β while dependent on l^*/h_o significantly, as expected.

Moreover, the observed scour hole profiles along the channel centreline were analysed. After several attempts, a moderate data scattering was achieved by normalizing the scour depth along the channel centreline with z^* and its location from the upstream end of the contraction model with l^*. The following tentative equation was found

$$Z = 2.92X^4 - 11.85X^3 + 15.36X^2 - 6.11X - 0.27, \tag{7}$$

where, with reference to the generic scour depth, z, along the channel centreline, Z is the ratio of z to the maximum scour depth z^* and X is the ratio of the distance of z from the upstream end of the contraction model to l^*. Eqn (7) needs a more extensive validation and then its coefficients cannot be considered as final, however it is interesting to observe how it predicts the scour hole and the sediment mound along the channel centreline confidently. To this purpose, Fig. 3 shows the comparison between the observed data and eqn (7) for runs 20, 21, and 22. In applying eqn (7), z^* was estimated from eqn (5) and l^* from eqn (6). Runs 20 and 21 were characterised by $Q = 0.040$ m^3/s, $h_o = 0.10$ m, $\beta = 0.70$, and $l = 3.00$ m and run 22 by $Q = 0.060$ m^3/s, $h_o = 0.15$ m, $\beta = 0.70$, and $l = 3.00$ m. It should also be noted that the bed degradation extends (downstream) beyond the contracted reach up to a distance of about $1.5 \cdot l$ in case of runs 20 and 22.

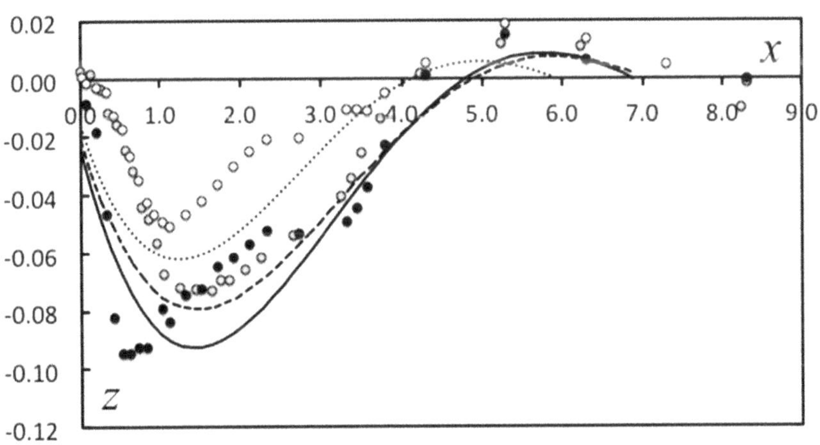

Figure 3: Comparison between observed and calculated values of z (in meters) along the channel centreline. x (in meters) is the distance of the point where z occurs from the upstream end ($x = 0$) of the contraction model. The grey symbols and the dashed curve refer to the run 20 at the time $t = 8$ hours, white symbols and dotted curve to the run 21 at the time $t = 1$ hour, and black symbols and full curve to the run 22 at the time $t = 21$ hours.

4 CONCLUSIONS

Laboratory experiments on spatial and temporal variations of the bed morphology in and around river contractions were carried out at University of Basilicata, Italy. Typically, runs lasted some days to achieve well-developed bed forms, but runs of short duration were also performed to acquire reliable data at the earlier stages of the bed forms development.

Experimental observations revealed that the magnitude of the scour hole would increase with increasing contraction length, and this aspect has not attracted much attention until now. In general, a single thalweg line develops along the channel axis, and it may extend downstream much farther than the contracted region.

Straightforward equations to predict the temporal development of the longitudinal thalweg profile were suggested in terms of the governing parameters. The main dimensionless parameters controlling the maximum relative thalweg depth are in order of importance: the contraction ratio (β), the relative length of the contracted reach (l/h_o), the approach densimetric Froude number (F_d), the dimensionless time (T), and the relative submergence (h_o/d_{50}). The relative length of the thalweg profile is mainly dependent on l/h_o and F_d. The ranges of validity for the proposed equations (i.e. eqns (5) and (6)) are $18.81 \leq h_o/d_{50} \leq 88.65$, $4.94 \leq l/h_o \leq 30.00$, $0.70 \leq \beta \leq 0.90$, $1.08 \leq F_d \leq 2.44$, and $1.0 \cdot 10^3 \leq T \leq 2.1 \cdot 10^6$. These ranges seem all to get wider when considering literature data.

Further research studies might concern the applicability of the proposed time-dependent relationships under unsteady flow regimes, and the investigation of live-bed conditions.

REFERENCES

[1] Straub, L.G., Effect of channel-contraction works upon regime of movable bed-streams. *Eos, Transactions, American Geophysical Union*, **15**(2), pp. 454–463, 1934.

[2] Gill, M.A., Bed erosion in rectangular long contraction. *Journal of the Hydraulics Division*, **107**(3), pp. 273–284, 1981.

[3] Dey, S. & Raikar, R.V., Scour in long contractions. *Journal of Hydraulic Engineering*, **131**(12), pp. 1036–1049, 2005.

[4] Laursen, E.M., Scour at bridge crossings. *Journal of the Hydraulics Division*, **86**(2), pp. 39–54, 1960.

[5] Komura, S., Equilibrium depth of scour in long constrictions. *Journal of the Hydraulics Division*, **92**(5), pp. 17–37, 1966.

[6] Lim, S.-Y., Clear water scour in long contractions. *Proceedings of the Institution of Civil Engineers - Water, Maritime and Energy*, **101**(2), pp. 93–98, 1993.

[7] Lim, S.-Y. & Cheng, N.-S., Scouring in long contractions. *Journal of Irrigation and Drainage Engineering*, **124**(5), pp. 258–261, 1998.

[8] Dey, S. & Raikar, R.V., Live-bed scour in long contractions. *International Journal of Sediment Research*, **21**(2), pp. 166–170, 2006.

[9] Brandimarte, L., D'Odorico, P. & Montanari, A., A probabilistic approach to the analysis of contraction scour. *Journal of Hydraulic Research*, **44**(5), pp. 654–662, 2006.

[10] Benedict, S.T., Clear-water abutment and contraction scour in the coastal plain and Piedmont provinces of South Carolina, 1996–1999. U.S. Geological Survey, Water-Resources Investigations Report 03-4064, Columbia, South Carolina, 2016.

[11] Mueller, D.S. & Wagner, C.R., Field observations and evaluations of streambed scour at bridges. U.S. Geological Survey, Water Resources Divisions Report No FHWA-RD-03-052, Louisville, Kentucky, 2005.

[12] Hong, S.H. & Abid, I., Physical model study of bridge contraction scour. *KSCE Journal of Civil Engineering*, **20**(6), pp. 2578–2585, 2016.

[13] Oliveto, G. & Marino, M.C., Spatial and temporal variations in bed morphology at river contractions. *Proceedings of the International Conference on Fluvial Hydraulics (River Flow 2014)*, pp. 1469–1474, 2014.
[14] Oliveto, G. & Marino, M.C., Experimental and numerical investigation of the bed morphology evolution in river contractions. *Proceedings of the 36th IAHR World Congress*, pp. 1697–1703, 2015.
[15] Oliveto, G. & Marino, M.C., Bed morphology changes at river contractions. *Proceedings of the 8th International Conference on Scour and Erosion*, pp. 855–860, 2016.
[16] Graf, W.H. & Altinakar, M.S., *Fluvial Hydraulics - Flow and Transport Processes in Channels of Simple Geometry*, John Wiley: Chichester, pp. 10–12, 1998.
[17] Oliveto, G. & Hager, W.H., Temporal evolution of clear-water pier and abutment scour. *Journal of Hydraulic Engineering*, **128**(9), pp. 811–820, 2002.

EFFECTS OF TEMPORAL VARIATION IN SEDIMENT REDUCTION FOLLOWING IMPROVED LAND MANAGEMENT PRACTICES ON END-OF-SYSTEM DELIVERY: A MODELLING INVESTIGATION OF A GRAZED CATCHMENT IN QUEENSLAND, AUSTRALIA

MELANIE E. ROBERTS[1,*] & ROBIN ELLIS[2]
[1]Australian Rivers Institute – Griffith University, Australia
[2]Queensland Government Department of Environment and Science, Australia

ABSTRACT

The world heritage listed Great Barrier Reef (GBR) is under threat. After climate change, water quality is recognised as the greatest stress on the reef. Sediment eroded from the catchments is transported into the reef lagoon, contributing to poor water quality. Poor water quality is linked to loss of habitat, coral death, reduced coral recruitment, algal blooms, and Crown of Thorns starfish outbreaks. The Queensland Government, in collaboration with the Australian Government under the joint Reef 2050 Water Quality Improvement Plan, undertakes a regular exercise to report on the health of the GBR and to track progress across a broad range of metrics, including water quality. This exercise comprises a combination of modelling and monitoring activities. Queensland Government periodically review their modelling framework and have identified areas for model improvement. One area identified is how the effects of changed land management practices are modelled. Presently, characteristics of recovery are empirically modelled, however parameterisation is challenged by a sparsity of data. The temporal characteristics of recovery are not presently explicitly accounted for within the modelling process. This study explores the variation in modelled end-of-system sediment loads for an exemplar sub-catchment as a result of employing varying temporal signatures of recovery for grazed lands. The purpose of this study is to identify whether the modelled end-of-system sediment loads are sensitive to the time-signature of recovery, to inform how land management practices are included in the models. The results show that the time-signature of recovery can have significant effects on the end-of-system sediment loads where changes in land management practices coincide with local hot-spots of sediment generation. Consideration of the temporal dynamics of erosion reduction due to different intervention strategies is essential to provide the best opportunity to meet water quality targets in the GBR lagoon.
Keywords: sediment, catchment, re-vegetation, gully, source, Great Barrier Reef.

1 INTRODUCTION

The Great Barrier Reef (GBR) is under threat [1], [2]. Climate change is placing unprecedented pressure on the GBR [3], leading to coral bleaching [4] and more severe cyclonic activity [5]. In addition to tackling climate change, other pressures on the GBR need to be reduced, to minimise damage and promote recovery [6], [7]. Avoiding synergistic impacts from multiple stressors is key to preventing rapid decline of reef health, while ensuring greater resilience of the reef to current and future stressors [8].

Sediment and nutrients transported from catchments in the GBR significantly impact water quality and ultimately the health and recovery of the reef [6]. Sediment affects the reef health through a variety of mechanisms. Suspended sediments reduce light availability, which impedes seagrass growth [9], an important habitat and food supply for reef species such as dugong [10]. Reduced light availability also affects the coral's algal symbiont, ultimately

* *ORCID: http://orcid.org/0000-0003-4027-9651*

leading to coral death. Sedimentation can bury coral polyps, cause tissue necrosis, and reduce the recruitment and survival of coral larvae [11]. Sediment also assists the transportation of nutrients from the catchments, potentially contributing to eutrophication [12]. High nutrient loads stress vulnerable coral [13], making them more prone to bleaching, and are linked to algal blooms and Crown of Thorns starfish outbreaks [14].

Improving water quality in the GBR lagoon is a joint priority for the Australian and Queensland Governments. Implementation of the Reef 2050 Water Quality Improvement Plan 2017–2022 aims to reduce anthropogenic end-of-catchment loads of fine sediment by 25%, particulate nutrient loads by 20%, and dissolved nitrogen by 60% [15]. To meet these targets, action is required to reduce sediment erosion and nutrient runoff. The 2017 Scientific Consensus Statement [13] indicates that current activities are insufficient to meet the identified targets.

The next few years will be critical in the fight to save the GBR. In 2015 and 2016 the GBR experienced back-to-back bleaching events. Reducing sediment-linked stressors is critical to allow the reef to recover from previous events. Actions that reduce sediment and nutrient levels in the reef lagoon in the long run may be inadequate to assist the GBR if they do not deliver significant short term benefits.

Gully erosion is a significant component of sediment transported to the GBR [16]. Gully erosion is responsible for 28% of sediment transported to the GBR World Heritage Area [17], despite occupying a small fraction of the landscape. The remaining sediment is due to hillslope erosion (63%), and streambank erosion (9%) [17]. Due to the high contribution of gullies to the total sediment budget, gully remediation and repair has been a focal point for research [18] and investment [19]. While it is known that land use practices such as managing stocking rates will reduce erosion, by reducing land disturbance and allowing ground cover rates to increase and mature trees to establish, what is not well understood is how long it takes for changes in land management practices to translate into observed reductions in the sediment load. As the protection of the GBR is time-critical, it is important to know what the expected benefit of different interventions would be over the short term, not just in the long-term once the full benefits are realised.

The Queensland Government, in collaboration with the Australian Government, undertakes regular modelling exercises to assess the expected outcomes resulting from changed land use and land management practices on the GBR through the P2R (or paddock-to-reef) program. The results of this exercise are communicated through the GBR Report Cards, which provides an evaluation of the expected long-term average sediment and nutrient levels within the reef lagoon resulting from current, or projected, land use practices. A long-term average is obtained by modelling sediment and nutrient transport over a 28-year climate scenario that includes a range of climatic conditions. Although there is considerable uncertainty in these scenarios, this approach nevertheless provides insight into the expected outcomes of changes in land use and land management practices in the long-term. These insights must be evaluated with the understanding that as erosion and nutrient run-off are predominantly rain-driven events, years of low rainfall will yield lower sediment and nutrient levels, while years of high rainfall will yield higher sediment and nutrient levels in the GBR lagoon.

Critically, this modelling approach, focussed on the long-term average, does not consider the temporal trajectory of recovery. That is, the models are applied under the assumption that the full benefits (or losses) due to changes in land use or land management practice have been realised, irrespective of how long it would take for this to occur. This approach is appropriate to the purposes of the GBR Report Cards. It is important to emphasise that implicitly considering temporal changes in environmental performance of the landscape as a function of changing land management practice is not directly relevant to the P2R loads reporting program. The P2R program reports the average annual constituent load based on the status quo landscape management scenario. If however the primary interest were on the short-term impacts or land management practice change, use of the modelling outputs for the GBR Report Cards as the only guide to investment for gully remediation could be problematic as differences between interventions are not explicit in the modelling process, and therefore do not provide an indication of the expected reductions in sediment load in the short term. Although not developed to assess short-term or transitory effects, the models and associated data brought together under the P2R program, appropriately modified, nevertheless provides an opportunity to explore such short-term effects.

While further research is required to identify the specific trajectories of erosion reduction due to different interventions, we can nevertheless investigate the relative effects of temporal variation in sediment reduction due to improved land management practices from a theoretical perspective. Using a grazed catchment in the Burdekin as a case study, we apply different temporal recovery rates to the management factor in the gully erosion equations that represents land management practices (see Section 2.1 for a description of the model) to investigate how this variation affects the end-of-system sediment delivery.

This research demonstrates that the form of the temporal variation assumed is important, with end-of-system delivery dominated by the interaction of the management factor and the rainfall. To obtain a reliable indicator of the short-term benefits of improving land management practices, this research has shown that the temporal trajectory of recovery must be considered.

This paper is organised as follows. The Methods section (Section 2) provides an overview of the modelling framework in Section 2.1 and introduces the case study catchment in Section 2.2. The results are presented in Section 3 together with a discussion in Section 4. A short conclusion is provided in Section 5.

2 METHODS

2.1 Modelling framework and model for gully erosion

Queensland Government uses the eWater Source [20] platform to model catchment processes for the GBR Report Cards including the processes of sediment erosion, deposition and transport, which are the focus of this analysis. For our analysis we use an instance of the model used for the GBR Report Cards, with modifications to the gully erosion parameters as detailed below. A complete description of the modelling framework for the GBR Report Cards is not provided, but can be found in [21], as we confine our discussion to those aspects of the modelling relevant to our study, namely erosion from gullies. We note however that the eWater Source platform allows for spatial variation in the landscape to be represented in the models. A Functional Unit (FU) approach is employed, where the landscape is divided into units that reflect different land uses, e.g. open grazing or sugar cane. Within each FU a single parameter set is employed, and thus the parameters for each unit represent a summary of the variation within that unit.

The Dynamic SedNet model is used to simulate the supply of sediment to the river system from the erosion of gullies, hillslope and streambanks. For a complete description of the Dynamic SedNet model see [22] and [23]. Erosion from gullies is calculated by relating rainfall driven runoff to the volume of gullies exposed to that runoff, according to eqns (1)–(5). The long-term annual average sediment supply, *AASS* [t/year] from each FU is given by the product of the total volume of gullies within the functional unit, V [m^3], and the soil dry bulk density P_s [$\mathrm{t/m}^3$], divided by the Age [years] of the gully, see eqn (1). The daily sediment load [kg/day] is calculated by weighting the annual average sediment supply by the ratio of the daily runoff, $Q_j(t)$ [mm/day], to the long term average runoff, Q_T [mm/year]. The daily sediment load is apportioned into a fine component, *DFSL* eqn (2), and a coarse component, *DCSL* eqn (3), using the proportion of fine sediment, θ. Fine sediment is defined to be $\leq 20\mu\mathrm{m}$. The factor 1000 converts the sediment yield from tonnes to kg. We explicitly indicate the temporally varying terms using (t) notation:

$$AASS = \frac{P_s V}{\text{Age}}, \tag{1}$$

$$DFSL(t) = 1000\, AASS\, (Q_j(t)/Q_T)\, \theta, \tag{2}$$

$$DCSL(t) = 1000\, AASS\, (Q_j/Q_T)\, (1-\theta). \tag{3}$$

The *DFSL* eqn (2) is further adjusted to account for the impact of land management practices on erosion through the dimensionless management factor M_f^*. Improved land use management practices decrease the daily fine sediment load, $M_f^* < 1$, while poor practices amplify the erosion, $M_f^* > 1$. A further term, the Activity Factor, enables gullies to be switched off when they cease to be productive. As our case study deals only with productive gullies, and therefore Activity Factor $= 1$, we neglect this term. Erosion of coarse sediment is assumed to be unaffected by management activities, and therefore is not adjusted from eqn (3). Moreover, as sediment transport within the stream network accounts only for fine sediment, end-of-system sediment loads are determined from the erosion and deposition of fine sediment only. The daily fine sediment yield from gullies for each functional unit that can be transported through the stream network, $S_{FU}^*(t)$ [kg/day] is therefore

$$S_{FU}^*(t) = 1000 \frac{P_s V \theta M_f^*}{\text{Age}} \frac{Q_j(t)}{Q_T}. \tag{4}$$

The $*$ notation indicates that the daily sediment yield relates to a time-invariant management factor. As detailed information about the year of gully initiation is not known, it is assumed that gully initiation coincided with the commencement of intensive grazing, and that Age $=$ 100 for all functional units within the case study.

Eqn (4) calculates the daily fine sediment yield from gullies within a functional unit as a function of the daily runoff, soil characteristics, gully volume and age, modified by the management factor to account for the effects of land management practices on the realised erosion. Land management practices captured through M_f^* include stocking rates, re-vegetation of degraded land, and engineering solutions such as rock capping gully heads, or the installation of porous check dams. Engineering solutions that effectively eradicate the gully are better captured through the reduction in the gully volume parameter, V, such solutions are not considered in this analysis.

To investigate the effects of temporal variation in sediment reduction resulting from improvements to land management practices, we modify eqn (4) to introduce a time-dependency in the management factor M_f^*. As the rate of sediment reduction due to different land management practices is not currently known, we explore the effects of temporal variation in the reduction by considering some exemplar rates. We update eqn (4) to reflect a time-varying management factor, $M_f(t)$. Thus the daily sediment yield calculated with a time-dependent management factor, $S_{FU}(t)$ [kg/day], is given by

$$S_{FU}(t) = 1000 \frac{P_s V \theta M_f(t)}{\text{Age}} \frac{Q_j(t)}{Q_T}. \tag{5}$$

The GBR Report Cards consider four classes of land management practice, rated D to A, and relate these classes to the management factor according to Table 1. The 2016 Report Card indicates that 25% (7 599 458 ha) of gully area was managed under best management practice systems, that is A-Class. Incorporating pastures and streambanks, 36% of grazing

Table 1: Relative gully erosion rate by management practice class for grazed land, adapted from [25, Table 11] (originally adapted from [26, Table 4]). [25] define land management practices for grazing under the ABCD framework as A (highly likely to maintain land in good condition), B (likely to maintain land in good/fair condition), C (likely to degrade some land), and D (highly likely to degrade land to poor condition).

Grazing practice management class	A	B	C	D
Management factor M_f^*	0.75	0.9	1	1.25

lands in the GBR catchments were managed under best practice, requiring considerable improvement to reach the 2018 target of 90% of grazing lands managed in accordance with A-class practices [24].

To explore the effect of a time-varying management factor on end-of-system delivery we consider four different rates of erosion reduction under six timeframes for recovery, a total of 24 scenarios. In all cases we assume that the simulation commences with gully erosion in accordance with C-class management, that is $M_f(0) = 1$, and by the end of the 28 year simulation period 90% of gully area has achieved the erosion reductions associated with A-class management, that is $M_f(t_{final}) = M_f^\dagger = 0.775$. We consider four different rates of improvement, a step function that shows no benefit until the year of full benefit is reached, as well as linear, quadratic and exponential declines in erosion. The six different timeframes correspond to increasing time required for the full benefits of management practice change to be achieved. This analysis compares recovery periods of $\tau = 2, 5, 10, 15, 20$ and 25 years. These are compared with the baseline case of full benefit achieved throughout the simulation, that is $M_f(t) = 0.775$. Refer to Table 2 for the equations used to form the time-dependent management functions $M_f(t)$. Fig. 1 shows the form of the management factor time series for three of the recovery periods considered together with the baseline case. The management factor is updated annually, at the start of each simulation year (1 July), therefore within each simulation year the management factor is held constant. The 28-year simulation period is 1 July 1986 to 30 June 2014, which includes a variety of weather events including periods of high and low rainfall. We note that neither this analysis, nor the P2R program, seeks to model historical loads associated with this time period.

2.2 Study site

The Bowen catchment is located within the greater Burdekin catchment, a 146,000 km^2 area that drains into the GBR south of Townsville in far north Queensland, as shown in Fig. 2. The Burdekin catchment is the largest source of contemporary sediment to the GBR lagoon [27]. The Burdekin catchment is dominated by grazing, which occupies 91% of the region [27], and has been linked to vegetation clearing and an acceleration in sediment supply to the GBR. Cattle numbers have increased from <0.05 million in 1860 to 1.3 million in 2016–17 [28]. The Bowen catchment occupies 7% of the Burdekin catchment and discharges below the Burdekin Falls Dam. An estimated 94.7% of sediment reaching the headwaters of the Bowen catchment will be discharged into the GBR lagoon.

3 RESULTS

The effects of varying the gully management factor on the end-of-system delivery of fine sediment are shown in Figs 3–5. The results are presented as excess sediment delivered to

Table 2: Equations for the annual management factor $M_f(t)$ during the recovery period for each of our studies, where $t = \tau$ is the year in which the full benefits of changed land practices are achieved, $M_f(0)$ is the initial management factor at the start of the simulation period and $M_f^\dagger$ is the final management factor achieved due to changed land practices. Once the full benefits of recovery are achieved $M_f(t) = M_f^\dagger$, $t > \tau$.

Recovery form	**Management factor $M_f(t)$**
Step	$M_f(0) \quad t \leq \tau$
Linear	$\frac{M_f^\dagger - M_f(0)}{\tau} t + M_f(0) \quad t \leq \tau$
Quadratic	$\frac{M_f^\dagger - M_f(0)}{\tau^2} t^2 + M_f(0) \quad t \leq \tau$
Exponential	$\left[M_f^\dagger - M_f(0)\right] \left[\frac{\exp(-0.25t) - 1}{\exp(-0.25\tau) - 1}\right] + M_f(0) \quad t \leq \tau.$

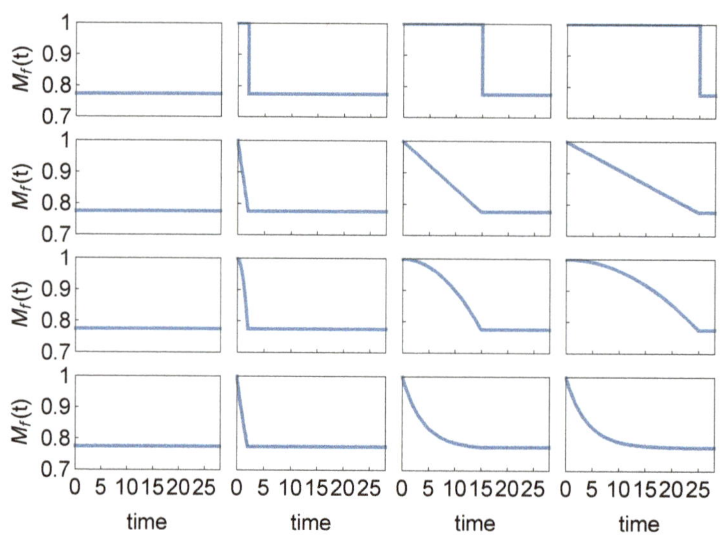

Figure 1: Exemplar time-dependent management factor, $M_f(t)$, values used for this analysis. Each column represents a different recovery period. The baseline case of instantaneous recovery is shown in column 1, column 2 corresponds to a 2-year recovery period, column 3 to a 15-year recovery period, with column 4 being a 25-year recovery period. The analysis also considers recovery periods of 10 and 20 years. The rows correspond to the four assumed rates of recovery, namely step, linear, quadratic and exponential. Refer to Table 2 for the corresponding equations.

Figure 2: Location of the Bowen catchment in the state of Queensland.

the GBR lagoon, we refer to this as the delivered excess sediment. The delivered excess sediment is calculated by scaling the original sediment load, in kilotonnes [kt], by the River System Delivery Ratio, and subtracting the baseline comparison case. The River System Delivery Ratio represents the proportion of sediment from a catchment that will be delivered to the GBR lagoon. Unless a catchment discharges directly into the lagoon, some of the fine sediment discharged from a catchment will be deposited before reaching the lagoon. Previous analysis has identified the River System Delivery Ratio for the Bowen catchment to be 0.947113. The baseline comparison case, which reflects immediate full benefits from 90% of grazing lands managed by A-class practices, is subtracted from the sediment load to better illustrate the effect of improved land use management practices being delayed. Once a scenario achieves the full benefits of recovery, that is $M_f(t) = 0.775$, the sediment load will be equal to that of the baseline case, and the excess sediment goes to zero.

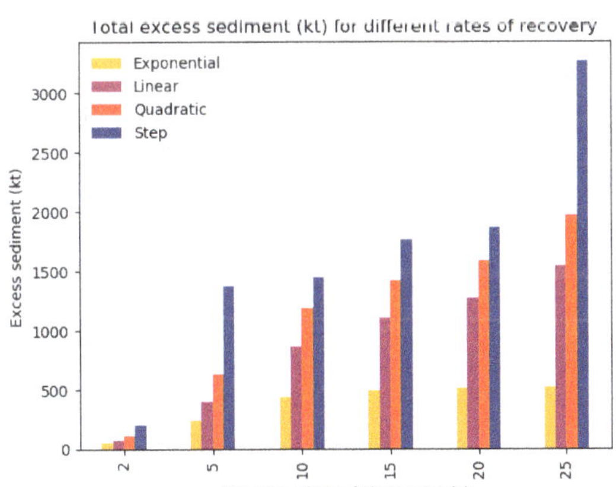

Figure 3: Sum of the excess sediment exported to the GBR lagoon over the 28-year simulation for the 24 assumed management factor time-series, grouped by the recovery period.

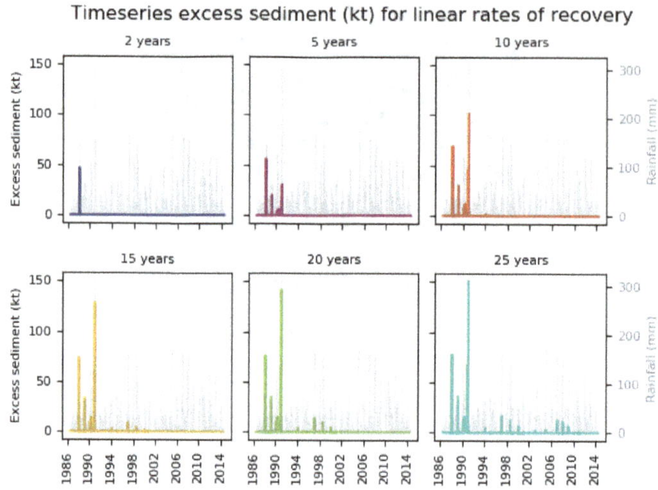

Figure 4: Timeseries of excess sediment [kilotonnes] delivered to the GBR lagoon under linear recovery regimes for the six different τ scenarios. Daily rainfall [mm] is provided for reference.

As detailed in Section 2, a 28-year simulation period is employed for all simulations, using historical weather observations from 1 July 1986 to 30 June 2014. The total delivered excess sediment over the simulation period for the 24 scenarios is shown in Fig. 3.

The time series of delivered excess sediment for the 6 recovery periods with a linear recovery form is shown in Fig. 4. The cumulative delivered excess sediment for the 24 scenarios is shown in Fig. 5. Fig. 5(a) arranges this data by the rate of recovery, while Fig. 5(b) shows this same data arranged by the recovery period. Daily rainfall [mm] from Bureau of Meteorology station 33176 (Gatton Vale), located within the Bowen catchment, is shown for reference in Figs 5 and 4, as gully erosion varies significantly between wet and dry seasons.

4 DISCUSSION

This analysis indicates that the assumed form of recovery due to improved land use management practices significantly affects the modelled sediment load delivered to the GBR lagoon. Both the rate of recovery (step, linear, quadratic or exponential) and the timeframe to achieve the full benefits of recovery (indicated by the recovery period τ) are shown to be important. Across the full 28-year simulation, shifting 90% of grazing lands from C-class to A-class management results in a 12.4% reduction in fine sediment reaching the GBR lagoon. In comparison, a delay of 5 years before seeing any benefits from improved practices decreases this benefit to no more than 7.5% (as modelled by the step function). To place this in perspective, 60% compliance with A-class practices throughout the simulation ($M_f(t) = 0.85$) corresponds to an 8.3% reduction in total exported fine sediment. Therefore a 5 year delay in achieving erosion reduction is equivalent across the simulation period to missing the target for area of land managed according to A-class practices by more than a third. A delay in the reduction in erosion following improved practices of this order is not infeasible. While there is a lack of published research on the temporal evolution of erosion reduction following natural regeneration or revegetation of grazing lands, in a different context [29] found dense

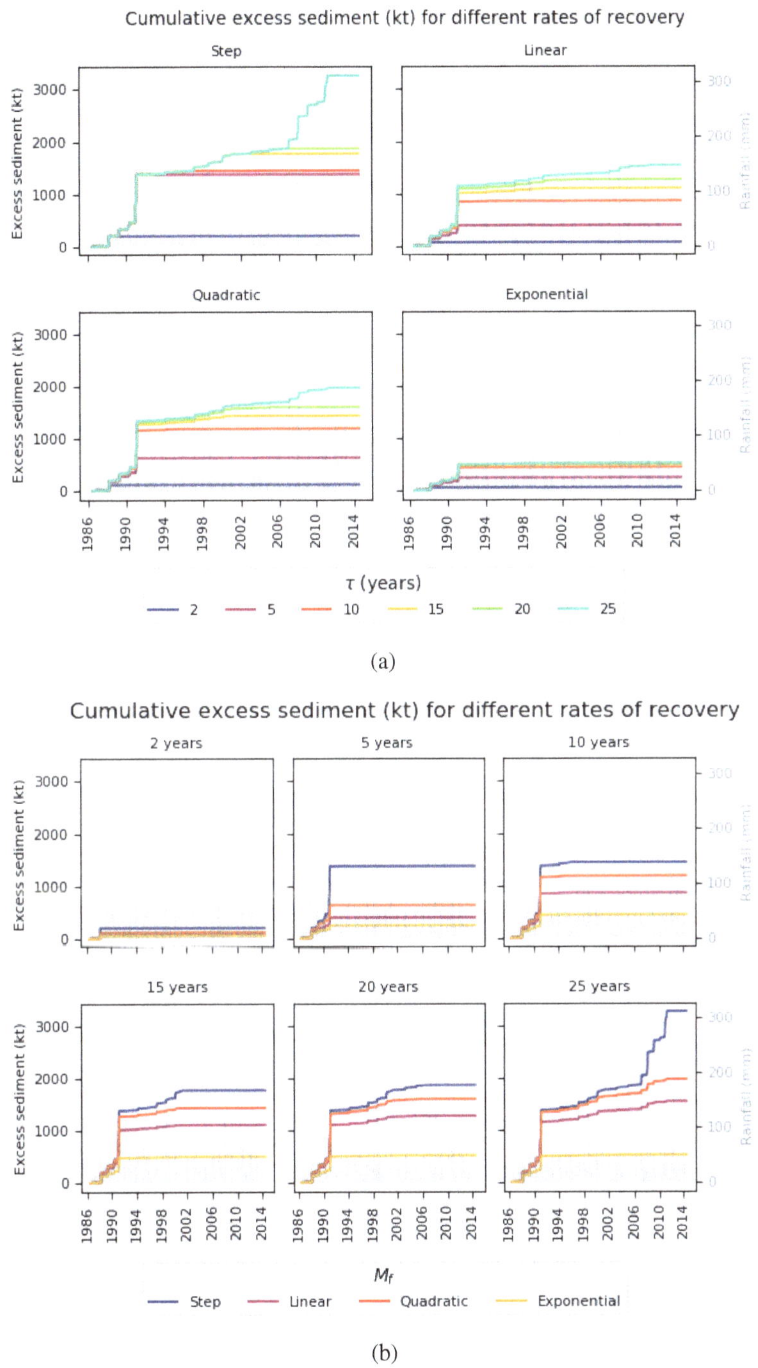

Figure 5: Cumulative excess sediment [kilotonnes] delivered to the GBR lagoon for the 24 assumed management factor time-series. The upper plot (a) groups the results by the form of recovery, while the lower plot (b) groups the same results by recovery period. Daily rainfall [mm] is provided for reference.

canopies typically required 10 years to develop following revegetation, illustrating the time lag between management action and environmental benefits.

The excess sediment delivered to the GBR under the different scenarios is not uniform in time. Gully erosion is a rainfall driven event, and therefore natural variation in the rainfall will result in variation in erosion. Furthermore, our analysis considers temporal variation in the rate of erosion as a model for delays in the benefits of management action. The impact of sediment on the GBR cannot be reduced to a consideration of average sediment loads at the entry to the lagoon. Different organisms within the GBR will have individual tolerance thresholds for both the length and quantity of exposure. We must therefore consider the temporal trajectory of delivered sediment when evaluating the impact of changed land management practices on the GBR. While it is beyond the scope of this paper to link our catchment models to impacts on the GBR, we nevertheless explore how the rate of recovery potentially impacts the GBR. Fig. 4 shows the temporal trajectory of delivered excess sediment for the 6 scenarios with a linear recovery form. Excess sediment is displayed to better show differences between the recovery periods, peak events correspond to the high rainfall events shown. As expected, the magnitude of peak events is greater the longer the recovery period. It is therefore likely that throughout the recovery period acute stresses from peak erosion events will be further amplified due to the delay between management action and the benefits of reduced erosion. Comparable results were obtained for the step and quadratic forms, however the amplification of peak events under the exponential form was considerably reduced after approximately 10 years. This is due to the similarities in the exponential functions for $M_f(t)$ with $\tau \geq 10$, as shown in Fig. 1.

Variation in the excess sediment loads is driven by the intersection of the gully management factor and the rainfall. The impact of this intersection is well illustrated by the large rainfall events in April 1991. Occurring towards the end of year 5, this event has no effect on the excess sediment in the $\tau = 2$-year simulations (see Fig. 5b), as once the full benefits of improved practices are realised, $t > \tau$, the excess sediment exported to the lagoon is zero. In comparison, the $\tau = 5$-year simulations show significant differences between the different recovery forms. The exponential form, being close in value to the long-term objective of $M_f(t) = 0.775$, sees minimal impact from this large rainfall event, whereas the step function experiences a more than 5-fold increase in excess sediment over a short period of time. In contrast, the next 5-year period has relatively low rainfall resulting in modest increases in excess sediment for the step function simulations, as shown by comparing the 5 (pink) and 10 (orange) year lines for the step profile in Fig. 5(a). As erosion is rain-driven, variations in rainfall will lead to natural variation in erosion. Therefore, obtaining a realistic estimate of the temporal trajectory of fine sediment supply following remediation actions requires consideration of climate variation.

It is expected that different interventions will have potentially vastly different rates of recovery. These differences need to be considered when evaluating different intervention strategies. If a period of low rainfall follows intervention, the benefits to the GBR between different approaches may well be similar. However, as shown in Figs 4 and 5, where a period of high rainfall occurs soon after intervention, different approaches, and hence recovery rates, will result in markedly different sediment loads entering the GBR lagoon. This has the potential to result in significantly different outcomes for the affected areas of the GBR. Further research is required to identify which recovery rates are representative of different approaches, and the expected impacts on the GBR.

5 CONCLUSION

Reducing sediment-linked stressors on the GBR is essential to enable damaged reefs to recover and to promote resilience to climate change. These improvements are required in the short-term, therefore it is imperative that the temporal trajectory of recovery is considered when evaluating the expected benefits of different interventions to reduce erosion. A lag between gully remediation activities and achieving reduction in erosion is not unexpected, particularly for management actions such as reducing stocking rates, fencing disturbed land, and revegetation projects.

The modelling framework, and associated data, developed for the P2R program provides an opportunity to explore the potential effects of such a time lag through a simple modification to the gully management factor within the equation for gully erosion. This paper presents a theoretical investigation into the variation in fine sediment loads delivered to the GBR lagoon from a case study catchment due to delays in achieving erosion reduction from improved management practices. We demonstrate that both the rate of recovery and the recovery period, that is until full benefits of improved practices are achieved, significantly affects the load of sediment reaching the GBR lagoon. Consequently, evaluation of different intervention activities and targets requires knowledge of the temporal trajectory of recovery. Failure to consider the temporal trajectory of recovery could lead to remediation actions being identified that are insufficient to meet objectives for water quality.

Research is required to identify the appropriate recovery trajectories for different interventions. In the mean time the methods presented in this study, together with expert knowledge, may be used to form indicators of the likely sediment loads over time.

ACKNOWLEDGEMENT

This project has received funding support from the Queensland Government through the Queensland Water Modelling Network.

REFERENCES

[1] Hughes, T.P., Climate change, human impacts, and the resilience of coral reefs. *Science*, **301**(5635), pp. 929–933, Aug. 2003.

[2] Lewis, S.E. et al., Herbicides: A new threat to the Great Barrier Reef. *Environmental Pollution*, **157**(8-9), pp. 2470–2484, Aug. 2009.

[3] Hoegh-Guldberg, O. et al., Coral reefs under rapid climate change and ocean acidification. *Science*, **318**, pp. 1–7, Dec. 2007.

[4] Hoegh-Guldberg, O., Climate change, coral bleaching and the future of the world's coral reefs. *Marine and Freshwater Research*, **50**(8), pp. 839–29, 1999.

[5] Walsh, K.J.E. & Ryan, B.F., Tropical cyclone intensity increase near Australia as a result of climate change. *Journal of Climate*, **13**, pp. 3029–3036, Aug. 2000.

[6] Great Barrier Reef Marine Park Authority, Great Barrier Reef Outlook Report 2014, Tech. Rep., Jul. 2014.

[7] De'ath, G., Fabricius, K.E., Sweatman, H. & Puotinen, M., The 27-year decline of coral cover on the Great Barrier Reef and its causes. *Proceedings of the National Academy of Sciences*, **109**(44), pp. 17995–17999, Oct. 2012.

[8] Pandolfi, J.M., Connolly, S.R., Marshall, D.J. & Cohen, A.L., Projecting coral reef futures under global warming and ocean acidification. *Science*, **333**, pp. 1–6, 2011.

[9] Wooldridge, S.A., Preventable fine sediment export from the Burdekin River catchment reduces coastal seagrass abundance and increases dugong mortality within the Townsville region of the Great Barrier Reef, Australia. *Marine Pollution Bulletin*, **114**(2), pp. 671–678, Jan. 2017.

[10] Preen, A. & Marsh, H., Response of dugongs to large-scale loss of seagrass. *Wildlife Research*, **22**(4), pp. 507–514, 1995.

[11] Rogers, C.S., Responses of coral reefs and reef organisms to sedimentation. *Marine Ecology Progress Series*, **62**, pp. 185–202, 1990.

[12] Garzon-Garcia, A., Burton, J., Franklin, H.M., Moody, P.W., De Hayr, R.W. & Burford, M.A., Indicators of phytoplankton response to particulate nutrient bioavailability in fresh and marine waters of the Great Barrier Reef. *Science of the Total Environment*, **636**, pp. 1–12, Sep. 2018.
[13] Waterhouse, J. et al., 2017 Scientific Consensus Statement. Tech. Rep., State of Queensland, Aug. 2017.
[14] Brodie, J. et al., Synthesis of evidence to support the Scientific Consensus Statement on Water Quality in the Great Barrier Reef. Tech. Rep., State of Queensland, 2008.
[15] Eberhard, R. et al., Management options and their effectiveness. *Scientific Consensus Statement 2017*, Chapter 4, pp. 1–174, 2017.
[16] Bartley, R. et al., Sources of sediment, nutrients, pesticides and other pollutants to the Great Barrier Reef. *Scientific Consensus Statement 2017*, Chapter 2, pp. 1–105, 2017.
[17] McKergow, L.A., Prosser, I.P., Hughes, A.O. & Brodie, J., Sources of sediment to the Great Barrier Reef World Heritage Area. *Marine Pollution Bulletin*, **51**(1-4), pp. 200–211, Jan. 2005.
[18] Program, N.E.S., Project 3.1.7 reducing sediment loads to the great barrier reef: developing optimal approaches for treating alluvial gully erosion, https://nesptropical.edu.au/index.php/round-3-projects/project-3-1-7/. Accessed on: 18 Dec. 2018.
[19] Queensland Government, Great Barrier Reef Innovation Fund.
[20] eWater Source, https://ewater.org.au/products/ewater-source/.
[21] Australian and Queensland Governments, Paddock to Reef Integrated Monitoring, Modelling and Reporting Program. Tech. Rep., Nov. 2018.
[22] Ellis, R., Dynamic SedNet component model reference guide. Department of Science, Information Technology, Innovation and the Arts. State of Queensland, Tech. Rep., 2015.
[23] Wilkinson, S.N., Dougall, C., Kinsey-Henderson, A.E., Searle, R.D., Ellis, R.J. & Bartley, R., Development of a time-stepping sediment budget model for assessing land use impacts in large river basins. *Science of the Total Environment*, **468-469**(C), pp. 1210–1224, Jan. 2014.
[24] Australian and Queensland Governments, Great Barrier Reef Report Card 2016 Results – Reef Water Quality Protection Plan, pp. 1–155, 2017.
[25] Waters, D.K. et al., Modelling reductions of pollutant loads due to improved management practices in the great barrier reef catchments – whole of GBR. Queensland Department of Natural Resources and Mines, Toowoomba, Queensland, Tech. Rep., 2014.
[26] Thorburn, P.J. & Wilkinson, S.N., Conceptual frameworks for estimating the water quality benefits of improved agricultural management practices in large catchments. *Agriculture, Ecosystems and Environment*, **180**, pp. 192–209, Nov. 2013.
[27] Bartley, R., Croke, J., Bainbridge, Z.T., Austin, J.M. & Kuhnert, P.M., Combining contemporary and long-term erosion rates to target erosion hot-spots in the Great Barrier Reef, Australia. *Biochemical Pharmacology*, **10**(C), pp. 1–12, Jun. 2015.
[28] Australian Bureau of Statistics, 7121.0 Agricultural Commodities, Australia and state/territory and NRM regions, 2016-17. Excel spreadsheet, cat. no. 71210*DO*002_201617, 5 2018, http://www.abs.gov.au/AUSSTATS/abs@.nsf/DetailsPage/7121.02016-17?OpenDocument.
[29] Vesk, P.A., Nolan, R., Thomson, J.R., Dorrough, J.W. & Nally, R.M., Time lags in provision of habitat resources through revegetation. *Biological Conservation*, **141**(1), pp. 174–186, Jan. 2008.

FLUVIAL RESPONSES FOLLOWING VOLCANIC ERUPTIONS: THE BLANCO-ESTE RIVER, SOUTHERN CHILE

ANDRÉS IROUMÉ[1], VALERIA ZINGARETTI[1], DAMIÀ VERICAT[2], JOHNATHAN TENNY[3], MANEL LLENA[2] & RAMON J. BATALLA[2,4]
[1]Universidad Austral de Chile, Laboratory of Hydromorphology, Chile
[2]Fluvial Dynamics Research Group (RIUS), University of Lleida, Spain
[3]College of Forestry, Oregon State University, USA
[4]Catalan Institute for Water Research (ICRA), Spain

ABSTRACT

We examine the fluvial response of the River Blanco-Este (Chile) following the 1961 and 2015 eruptions of the Calbuco volcano. The river drains the north-eastern flanks of the Calbuco and was heavily affected by ash fall, pyroclastic flows and post-eruption channel reworking. The long-term morphological evolution of the river is analyzed along a 6.5 km-long segment using a sequence of remote aerial images. Since 2017, short-term fluvial responses are studied in two reaches representing the upstream and downstream parts of this segment, using multi-temporal high resolution topography and orthophotomaps obtained combining dGPS surveys and digital photogrammetry applied to photos acquired from drones, to monitor topographic changes in the channel topography and the area of exposed movable sediments, grain size and large wood deposits. Due to the lack of flow data we use photos from time-lapse cameras as a proxy to qualitatively characterize river flow. The active channel observed after the 1961 and 2015 eruptions has a similar pattern. According to the evolution of the channel morphology after the 1961 eruption, we hypothesize that the river reached a quasi-equilibrium condition in 2014, a situation severely modified by the 2015 eruption. Long-term (decadal) evolution of the river reflects the adjustment between sediment budget and channel morphology that progressively shifts from a braided pattern to a stable single-threat channel that illustrates the equilibrium between the water and sediments load, the grain-size and the gradient. The 2015 eruption increased sediment supply and modified channel morphology, a new braided configuration is observed. The annual to flood-based responses reflect the complexity of the geomorphic dynamics of the river: channel incision, development of single channel patterns and head cut erosion in the upper study reach, and continuous erosive-depositional changes of several meters thick, with evident incessant lateral migration and severe bank erosion in the downstream river reach.
Keywords: River Blanco-Este, channel morphologic evolution, volcanic eruptions, Chile.

1 INTRODUCTION

Natural disturbances such as volcanic eruptions, wildfires, landslides, debris flows, and floods can substantially modify hydrologic and geomorphic processes in watersheds by destroying or burying slope-stabilizing vegetation, by adding large volumes of sediments and large wood that can divert or dam stream channels, and by disrupting normal hydrologic pathways (interception, infiltration, and surface and subsurface runoff) [1]–[5]. The duration and severity of such disruptive effects vary with relief, climate, vegetation, and the degree and area of impact. Quantification of geomorphological changes and rates of landscape evolution is a matter of primary importance, as much in natural hazards studies [6] as in calibration of landscape evolution models [7]–[9]. It is thus important to study places where morphological changes are rapid enough to allow measurements to be made that reflect actual surface processes, while avoiding a large observation window which may smooth or even erase the short-term evolution [10].

WIT Transactions on Ecology and the Environment, Vol 234, © 2019 WIT Press
www.witpress.com, ISSN 1743-3541 (on-line)
doi:10.2495/RBM190031

Volcanic eruptions undoubtedly have the potential to inflict the largest impacts of all these disturbances in terms of scale and severity [5]. They can damage, destroy, or bury extensive areas of forest vegetation and cover the landscape with volcanic ash, filling river valleys, obliterating watersheds, disturbing drainage patterns and changing channel size, pattern and structure [5], [11]. Decay of tree roots compromises slope stability, dead trees contribute to large log jams on valley floors, and volcanic ash can wash off slopes for years following eruptions. Hydrologic, sedimentologic, and geomorphic responses to major explosive eruptions are dramatic, widespread, and persistent, and present enormous challenges to those entrusted with managing disturbance response. Volcanic eruptions have the potential to strongly alter river hydrology, as well as sediment and large wood dynamics, with consequences on human safety and infrastructures, and the environment [12]–[14].

Effects on the fluvial geomorphology and geologic and ecologic processes have been reported after eruptions in USA [15]–[17], Philippines [18], [19], Italy [20], Mexico [21], [22], New Zealand [23], Montserrat Island [24], and Perú [25]; i.e. Pierson and Major [5] presented a comprehensive analysis on the hydrogeomorphic effects of volcanic eruptions on drainage basins.

Very little was known in Chile on this issue until the Volcano Chaitén erupted in 2008; in fact, that eruption attracted researches worldwide whose works have enhanced the understanding of the short and medium-term effects of a volcanic eruption on adjacent river systems [11]–[14], [26]–[28]. Within this context, we have extended the research on the post-eruption fluvial responses studying the River Blanco-Este which drains the north-eastern flank of the Volcano Calbuco, following the last two eruptions in 1961 and 2015 to address the following research questions: a) To which extent the knowledge already acquired in river systems affected by the Chaitén volcano aids at better understanding of the fluvial response and recovery of the Blanco-Este? b) How do the intensity and type of volcanic disturbances (i.e. tephra fall, pyroclastic density currents (PDC) or dome collapses) control river morphology changes and subsequent recovery? c) Does the grain size of the sediments, for instance non cohesive sands in the Chaitén case and much coarser deposits in the Blanco-Este, control the evolution and recovery of the river channels? d) How does the sediments residence time in the drainage network inform on the rates of channel reworking and how does it in turn control the morphological evolution of the river? e) How different is the role of these controlling factors in distinct river systems and what is their relative weight on the evolution of the channel and its eventual recovery? and finally, f) How can river corridor management and the associated decision making processes be informed by the specific knowledge acquired?

Within this framework, here we present an initial study of: a) the long-term morphological evolution of the River Blanco-Este analysing sequences of remote images from 1961 to 2016 along a 6.5 km-long river segment, and b) the short-term fluvial responses in two reaches using HRT and orthophotomaps obtained combining dGPS surveys and digital photogrammetry applied to photos acquired from drones.

2 STUDY AREA

The study area corresponds to a 6.5 km-long channel segment of the River Blanco-Este (south Chile), that drains the north-eastern flanks of the Calbuco volcano and was heavily affected by ash fall, pyroclastic following the last two eruptions in 1961 and 2015. From the Calbuco crater at ≈1900 m a.s.l., the Blanco-Este flows for 13 km to join the River Hueñu-Hueñu and has an area of 39.6 km^2 at the confluence (see Fig. 1).

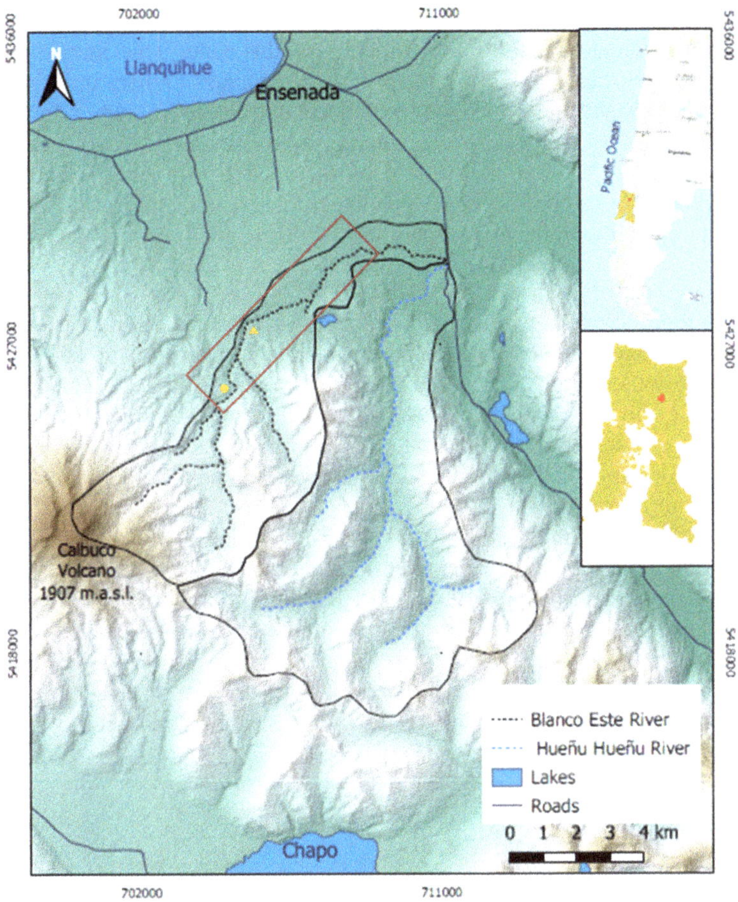

Figure 1: The study area, with the location of the River Blanco-Este draining the north-eastern flanks of the Volcano Calbuco. In a red rectangle, the 6.5 km-long channel segment. Yellow circle and triangle indicate the position of the upstream and downstream reaches.

The Calbuco is a stratovolcano localized in the Southern Andes (41°19'51"S, 72°36'46"W), which erupted on 22 April 2015 for the first time in 43 years, with very little warning. The Calbuco is one of the most active volcanoes in Chile and has a history of other twelve eruptions documented since the 18th century, occurring in 1792, 1893–1895, 1906–1907, 1911–1912, 1917, 1932, 1945,1961, 1972, with the 1893 and 2015 eruptions being the largest of this period [29], [30]. These events have included tephra fall, lahars, pyroclastic flows, and production of a dome and lava flows of basalt to basaltic-andesite composition. The 2015 event involved several phases during 22 and 23 April with the eruption column reaching a height of >15 km and delivering ca. 0.27 km^3 (dense rock equivalent) of basaltic-andesite tephra initially to the northeast and then east across Patagonia and ultimately circling the globe [30]. More than 40 cm of coarse-gravel tephra blanketed the terrain high on the cone and about 20 cm of tephra up to 2 cm diameter fell in the village of Ensenada. Pyroclastic flows were triggered by two processes: partial collapse of the eruption column and by the fall of hot, highly porous tephra (scoria, density ca. 1.5 g cm^{-3}) into the heads of steep drainages originating on the upper flanks of the cone. This immediately formed pyroclastic flows that raced down the steep, narrow channels. Where the tephra fell onto snow and ice fields, meltwater mixed with tephra and entrained alluvium to form cool lahars that moved rapidly down many channels radiating from the cone [29]–[31], as they have in other recent eruptions [32]. The northeast flank of Calbuco, where the Blanco-Este flows, received a higher proportion of hot flows than other areas around the cone. Fig. 1 shows the study segment together with the two reaches tackled in the analyses.

3 METHODS

The long-term morphological evolution of the 6.5 km-long study segment is analysed using a sequence of three aerial photos and remote sensing images from 1961 to 2016, a period that includes the last two eruptions (1961 and 2015). Historical photos and modern images are geo-rectified by comparing available LiDAR information and dGPS coordinates obtained at specific control points, in order to generate topographic models that will eventually aid at determining the long-term sediment budget of the river channel [33], [34].

In turn the short-term fluvial responses (months, years) are studied since 2017 in two reaches using HRT and orthophotomaps obtained combining Trimble® R5 PP/RTK dGPS surveys and digital photogrammetry applied to photos acquired from drones and analyses with Agisoft Photoscan Pro®. Multi-temporal HRT and orthophotomaps are used to study changes in channel topography and the area of exposed movable sediments, grain size and large wood deposits.

Due to the lack of flow data we use photos from time-lapse cameras as a proxy to qualitatively characterize river flow (i.e. no flow, low, medium and high flow).

4 RESULTS

4.1 Long term evolution

The long-term morphologic evolution of the River Blanco-Este was analyzed using a sequence of three remote images, one from 1961 (aero photo, Instituto Geográfico Militar, resolution 2.3 m) immediately after the 1961 eruption, and the others from 2014 (Bing, 5 m resolution) and 2016 (Google, 1–2.4 m resolution) just before and after the 2015 eruption, respectively. The first period 1961 to 2014 allows us to study how the river adjusted to the 1961 eruption, eventually reaching a *new equilibrium*. In terms of the second period, 2014–2015, we analyze how a single eruption abruptly altered channel's morphology and the overall valley floor configuration. After the 1961 eruption the mean active channel width in the study segment was 213 m, and by 2014 the channel had adjusted its morphology (i.e. single-threat meandering pattern) and showing a mean width of 78 m, altogether pointing to a *quasi-equilibrium* condition. The 2015 eruption severely modified this condition conveying the channel in the 6.5 km-long study segment to a mean width of 204 m and to a morphology almost identical to the one observed after the 1961 eruption (Fig. 2).

4.2 Short term evolution

In terms of the short-term effects of the 2015 eruption, field observations verify the extensive impact of this natural disturbance on the current morphology of the River Blanco-Este channel that the aerial images already displayed (Fig. 3).

Figure 2: From above to below, the active channel and water limits (black and blue lines, respectively) for year 1961, 2014 and 2016, and a 2016 image (Google®, 2 June 2016) of the study segment. Yellow arrows indicate flow direction.

Figure 3: The downstream study reach of the Blanco-Este before (left, DigitalGlobe®) and after (right, Google®) the 2015 eruption.

The upstream study reach of the River Blanco-Este valley displays channel incision, development of single channel patterns and head cut erosion. On the other hand, the downstream study reach shows massive accumulations of sediments that fill the valley floor offering a complex braided morphology, a multi-threat highly unstable pattern, with shifting channels and numerous bars that change their position on a flood basis. Sedimentation (of up to >3 m) appears to be the dominant processes in this valley reach, with erosion concentrating in the channel margins (Fig. 4). Both reaches are being quasi-permanently re-shaped by flush-floods that typically occur in the area (Fig. 5).

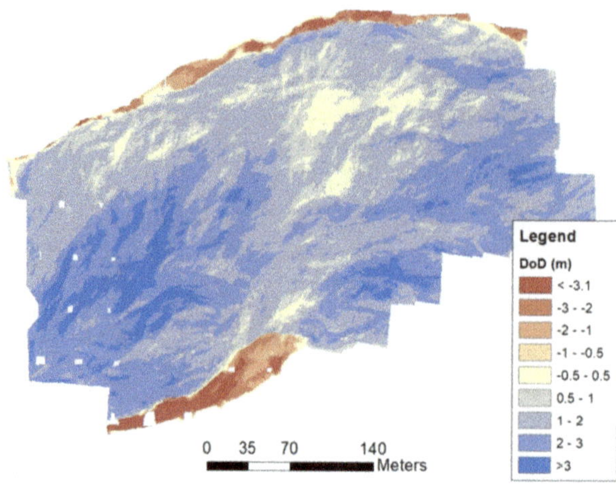

Figure 4: Raw difference of DEMs (DoD) between 11 April 2017 and 1 August 2017 on the downstream, study reach of the Blanco-Este (same spot as in Fig. 3). The colours displayed represent net change in elevation, expressed in meters. Negative values represent erosion (nearly entirely bank erosion in this case), while positive values represent deposition. The figure was produced using structure from motion photogrammetry processed in Agisoft Photoscan Pro® and GCD 6 for ArcGIS®. This DoD has not been classified to separate out vegetation or water and it has not been yet adjusted to account for uncertainty (i.e. minimum level of detection threshold of 0.5 m).

Figure 5: A view of flood dynamics in the downstream study reach of the Blanco-Este. Above from left to right, conditions at 16:00 on 18–20 July 2017. Below from left to right, conditions at 16:00 on 16–18 August 2017. Camera looking downstream.

Surface sediments are extremely poorly sorted, with large boulders, cobbles and gravels embedded in a sand matrix, conforming highly loose deposits easily mobilize by competent flows (Fig. 5), especially during austral winter and snowmelt related floods. Altogether, observations point out that the river is at the initial stages of a long-term readjustment process of the sediment budget after the perturbation. Downstream, at the confluence with the Hueñu-Hueñu, the Blanco-Este presents a rather stable gravel-bed morphology that leads us to hypothesize that the sediment wave affecting the valley after the eruption has not reached yet the lowermost segment of the river.

5 FINAL REMARKS

Based on evidences, we state that the river needed four decades to reach a more or less stable condition (or *quasi-equilibrium* condition) after the 1961 eruption, hence we hypothesize that the river system will require at least 40–50 years to reach again a stable morphological configuration. Nowadays, up-to-dated field and remote-sensing techniques allow tracking the short-term evolution of the new river channel and its responses through time, an approach that was not possible in the previous eruption period. So far, the river depicts extraordinary morphosedimentary activity following flood events, a fact that indicate that the alteration caused to valley floor was an intense and long-lasting phenomenon. Altogether, the Blanco-Este provides a worldwide opportunity to observe the medium to long long-term evolution of a fluvial landscape profoundly affected by the internal forces of the Earth further modified by the current hydroclimatic activity.

ACKNOWLEDGEMENTS

This research is developed within the Project FONDECYT 1170413 funded by the Chilean "Comisión Nacional de Investigación Científica y Tecnológica". Damià Vericat is currently a Serra Húnter Fellow at the University of Lleida.

REFERENCES

[1] Iverson, R.M., The physics of debris flows. *Reviews of Geophysics*, **35**(3), pp. 245–296, 1997.

[2] Moody, J.A. & Martin D.A., Initial hydrologic and geomorphic response following a wildfire in the Colorado Front Range. *Earth Surface Processes and Landforms*, **26**, pp. 1049–1070, 2001.

[3] Legleiter, C.J., Lawrence, R.L., Fonstad, M.A., Marcus, W.A. & Aspinall, R., Fluvial response a decade after wildfire in the northern Yellowstone ecosystem: A spatially explicit analysis. *Geomorphology*, **54**, pp. 119–136, 2003.

[4] Brummer, C.J. & Montgomery, D.R., Influence of coarse lag formation on the mechanics of sediment pulse dispersion in a mountain stream, Squire Creek, North Cascades, Washington, United States. *Water Resources Research*, **42**, W07412, 2006. DOI: 10.1029/2005WR004776.

[5] Pierson, T.C. & Major, J.J., Hydrogeomorphic effects of explosive volcanic eruptions on drainage basins. *Annual Review of Earth and Planetary Sciences*, **42**(1), pp. 469–507, 2014.

[6] Pasuto, A. & Soldati, M., An integrated approach for hazard assessment and mitigation of debris flows in the Italian Dolomites. *Geomorphology*, **61**(1–2), pp. 59–70, 2004.

[7] Hancock, G.R., Willgoose, G.R. & Evans, K.G., Testing of the SIBERIA landscape evolution model using the Tin Camp Creek, Northern Territory, Australia, field catchment. *Earth Surface Processes and Landforms*, **27**(2), pp. 125–143, 2002.

[8] Bogaart, P.W., Tucker, G.E. & deVries, J.J., Channel network morphology and sediment dynamics under alternating periglacial and temperate regimes: a numerical simulation study. *Geomorphology*, **54**(3), pp. 257–277, 2003.

[9] García-Castellanos, D., Vergés, J., Gaspar-Escribano, J. & Cloetingh, S., Interplay between tectonics, climate, and fluvial transport during the Cenozoic evolution of the Ebro basin (NE Iberia). *J. Geophysical Research*, **108**(B7), pp. 2347–2364, 2003.

[10] Garcin, M., Poisson, B. & Pouget, R., High rates of geomorphological processes in a tropical area: The Remparts River case study (Réunion Island, Indian Ocean). *Geomorphology*, **67**(3–4), pp. 335–350, 2005.

[11] Swanson, F.J., Jones, J.A., Crisafulli, C.M. & Lara, A., Effects of volcanic and hydrologic processes on forest vegetation: Chaitén Volcano, Chile. *Andean Geology*, **40**(2), pp. 359–391, 2013.

[12] Ulloa, H., Iroumé, A., Mao, L., Andreoli, A., Diez, S. & Lara, L.E., Use of remote imagery to analyse changes in morphology and longitudinal large wood distribution in the Blanco River after the 2008 Chaitén volcanic eruption, Southern Chile. *Geografiska Annaler: Series A, Physical Geography*, **97**(3), pp. 523–541, 2015. DOI: 10.1111/geoa.12091.

[13] Major, J.J. et al., Extraordinary sediment delivery and rapid geomorphic response following the 2008–2009 eruption of Chaitén Volcano, Chile. *Water Resources Research*, **52**(7), pp. 5075–5094, 2016. DOI: 10.1002/2015WR018250.

[14] Mohr, C.H., Korup, O., Ulloa, H. & Iroumé, A., Pyroclastic eruption boosts organic carbon fluxes into Patagonian fjords. *Global Biogeochemical Cycles*, **31**, pp. 1626–1638, 2017.

[15] Major, J.J., Post-eruption hydrology and sediment transport in volcanic river systems. *Water Resources Impact*, **5**(3), pp. 10–15, 2003.

[16] Dale, V.H., Swanson, F.J. & Crisafulli, C.M., *Ecological Responses to the 1980 Eruption of Mount St. Helens*, Springer-Verlag: New York, 2005.

[17] Major, J.J., Crisafulli, C.M., Frenzen, P. & Bishop, J., After the disaster: The hydrogeomorphic, ecological, and biological responses to the 1980 eruption of Mount St. Helens, Washington. *Volcanoes to Vineyards: Geologic Field Trips through the Dynamic Landscape of the Pacific Northwest: Geological Society of America Field Guide,* vol. 15, J.E. O'Connor, R.J. Dorsey & I.P. Madin (eds), pp. 111–134, 2009. DOI: 10.1130/2009.fld015(06).

[18] Hayes, S.K., Montgomery, D.R. & Newhall, C.G., Fluvial sediment transport and deposition following the 1991 eruption of Mount Pinatubo. *Geomorphology,* **45**, pp. 211–224, 2002.

[19] Gran, K.B. & Montgomery, D.R., Spatial and temporal patterns in fluvial recovery following volcanic eruptions: Channel response to basin-wide sediment loading at Mount Pinatubo, Philippines. *Geological Society of America Bulletin*, **117**(1–2), pp. 195–211, 2005.

[20] Branca, S., Geological and geomorphological evolution of the Etna volcano NE flank and relationships between lava flow invasions and erosional processes in the Alcantara Valley (Italy). *Geomorphology*, **53**(3–4), pp. 247–261, 2003.

[21] Inbar, M., Reyes, A. & Graniel, J.H., Morphological changes and erosion processes following the 1982 eruption of El Chichón volcano, Chiapas, Mexico. *Géomorphologie: Relief, Processus, Environnement*, 7(3), pp. 175–183, 2001.

[22] Tanarro, L.M., Andrés, N., Zamorano, J.J., Palacios, D. & Renschler, C.S., Geomorphological evolution of a fluvial channel after primary lahar deposition: Huiloac Gorge, Popocatépetl volcano (Mexico). *Geomorphology*, **122**(1–2), pp. 178–190, 2010.

[23] Manville, V. & Wilson, C.J.N., The 26.5 ka Oruanui eruption, New Zealand: A review of the roles of volcanism and climate in the post-eruptive sedimentary response. *New Zealand Journal of Geology and Geophysics*, **47**(3), pp. 525–547, 2004.
[24] Cole, P.D. et al., Deposits from dome-collapse and fountain-collapse pyroclastic flows at Soufrière Hills Volcano, Montserrat, in The Eruption of the Soufrière Hills Volcano, Montserrat From 1995 to 1999. *Geological Society London Memoirs*, **21**, pp. 231–262, 2002.
[25] Harpel, C.J., de Silva, S. & Salas, G., The 2 ka eruption of Misti volcano, southern Peru: The most recent Plinian eruption of Arequipa's iconic volcano. *Geological Society of America Special Papers*, **484**, pp. 1–72, 2011.
[26] Umazano, A.M., Melchor, R.N., Bedatou, E., Bellosi, E.S. & Krause, J.M., Fluvial response to sudden input of pyroclastic sediments during the 2008–2009 eruption of the Chaitén Volcano (Chile): The role of logjams. *Journal of South American Earth Sciences*, **54**, pp. 140–157, 2014.
[27] Ulloa, H. et al., Massive biomass flushing despite modest channel response in the Rayas River following the 2008 eruption of Chaitén volcano, Chile. *Geomorphology*, **250**, pp. 397–406, 2015. DOI: 10.1016/j.geomorph.2015.09.019.
[28] Ulloa, H. et al., Spatial analysis of the impacts of the Chaitén volcano eruption (Chile) in three fluvial systems. *Journal of South American Earth Sciences*, **69**, pp. 213–225, 2016.
[29] Mella, M. et al., Productos volcánicos, impactos y respuesta a la emergencia del ciclo eruptivo abril–mayo (2015) del volcán Calbuco. *Actas XVI Congreso Geológico Chileno*, pp. 98–101, 2015.
[30] Romero, J.E. et al., Eruption dynamics of the 22–23 April 2015 Calbuco Volcano (Southern Chile): Analyses of tephra fall deposits. *Journal of Volcanology and Geothermal Research*, **317**, pp. 15–29, 2016. DOI: 10.1016/j.jvolgeores.2016.02.027.
[31] Russell, A., Dussaillant, A., Meier, C., Rivera, A., Gonzalez, C. & Harrison, D., Controls on rapid post eruption fluvial system response, Calbuco, Chile. *Geophysical Research Abstracts, European Geophysical Union General Assembly*, **20**, EGU2018-9887, 2018.
[32] Moreno, H., Naranjo, J.A. & Clavero, J., Generación de lahares calientes en el Volcán Calbuco, Andes del Sur de Chile (41,3°S). *Actas XI Congreso Geológico Chileno*, pp. 513–516, 2006.
[33] Wheaton, J.M., Brasington, J., Darby, S.E. & Sear, D.A., Accounting for uncertainty in DEMs from repeat topographic surveys: improved sediment budgets. *Earth Surface Processes and Landforms*, **35**(2), pp. 136–156, 2010.
[34] Batalla, R.J., Iroumé, A., Hernández, M., Llena, M., Mazzorana, B. & Vericat, D., Recent geomorphological evolution of a natural river channel in a Mediterranean Chilean basin. *Geomorphology*, **303**, pp. 322–337, 2018.
DOI: 10.1016/j.geomorph.2017.12.006.

AREAL REDUCTION FACTORS FOR DESIGN RAINFALL ESTIMATION IN THE MODDER-RIET RIVER BASIN, SOUTH AFRICA

JACOBUS P. J. PIETERSEN & OCKERT J. GERICKE
Unit for Sustainable Water and Environment, Department of Civil Engineering,
Central University of Technology, South Africa

ABSTRACT

Design point rainfall estimates assume a uniform distribution of rainfall over a catchment and hence are only representative of a limited area. For larger areas, Areal Reduction Factors (ARFs) are used to convert design point rainfall depths or intensities to an average areal design rainfall depth or intensity for a catchment-specific critical storm duration and catchment area. This paper presents the development of an enhanced methodology to express the spatial and temporal rainfall variability at a quaternary catchment (QC) level by means of geographically-centred and probabilistically correct ARFs. The ARF values presented in this paper are based on observed daily rainfall data as extracted from 223 rainfall stations situated in the Modder-Riet River Basin (MRRB). The methodology adopted is based on a modified version of Bell's geographically-centred approach. Individual sets of ARF values were derived for each of the 23 QCs present in the MRRB by considering various storm durations and corresponding recurrence intervals. The differences in the regional sample ARF values highlight the presence of dominant weather types in each region and also confirm that ARFs are influenced by different rainfall-producing mechanisms, while not being constant for various storm durations and exceedance probabilities or recurrence intervals. It is recommended that the findings from this study and the use of geographically-centred probabilistically correct ARFs be expanded to other regions, both nationally and/or internationally to ultimately facilitate both improved design rainfall and flood estimation.

Keywords: Areal Reduction Factors, design point rainfall, areal rainfall, geographical-centred.

1 INTRODUCTION

Design point rainfall estimates are only representative of a limited area and for larger areas the areal average design rainfall depth or intensity is likely to be less than the maximum design point rainfall depths or intensities [1]. Areal Reduction Factors (ARFs) are used to describe the relationship between point and areal rainfall by converting design point rainfall depths or intensities to an average areal design rainfall depth or intensity for a catchment-specific critical storm duration and catchment area [2].

Flood-producing rainfall events are mostly characterised by a non-uniform spatial and temporal rainfall distribution. Typically, rainfall storms could have one or more maximum rainfall cores, while displaying for any given period a sensibly smooth non-linear reduction in average areal values with an increasing distance from the maximum rainfall core [2].

2 ARF ESTIMATION METHODS

In practice, engineers and/or hydrologists are in most cases concerned with design rainfall, i.e. rainfall information derived from observed rainfall data and which comprises of a depth and duration associated with a given recurrence interval (*T*) or annual exceedance probability (AEP) [3].

ARF estimation methods as applicable to design rainfall, are normally grouped into the following two broad categories.

www.witpress.com, ISSN 1743-3541 (on-line)
doi:10.2495/RBM190041

2.1 Analytical methods

Analytical methods are derived mathematical algorithms used to characterise the spatial and temporal rainfall variability by incorporating simplified assumptions that are not entirely true descriptions of the actual rainfall process [1], [4]. The fact that the actual rainfall processes are partially ignored is a cause for concern and this is further compounded by the often limited amount of actual rainfall data used during the verification of these methods. In response to these inherent shortcomings, several new analytical methods to estimate ARFs have been proposed during the last three decades such as storm movement [5], crossing properties [6], spatial correlation structure [7] and scaling relationships [8].

2.2 Empirical methods

Empirical methods can either be based on a geographically-centred or storm-centred approach. The geographically-centred approach describes the relationship between areal average design rainfall over a geographically fixed area (e.g. catchment area) and a corresponding design point rainfall value representative of the area under consideration. In other words, ARFs are used for percentage reduction, which relates to the statistics of point and areal design rainfall and considers the uniform temporal and spatial distribution of rainfall over a catchment area.

In the storm-centred approach, the estimation of areal design rainfall is not limited to a fixed geographical area, but rather associated with the extent of individual storm rainfall events and the way in which the rainfall intensity decreases with distance from the central maximum rainfall core [2], [4].

The empirical and/or analytical estimation of ARFs on a large scale is basically limited to the United Kingdom [9], United States of America [10], [11] and Australia [1]. Omolayo [12] also highlighted that, apart from these studies, not much research has been conducted in other parts of the world and ascribed this to insufficient rainfall-monitoring networks and a lack of short duration (sub-daily) rainfall data. Furthermore, the inconsistency present in the ARF results obtained from using the above data-intensive empirical and/or analytical methods internationally, is also a major concern and could be ascribed to the variation in predominant weather types, storm durations, seasonal factors and recurrence interval [13]–[15].

According to Svensson and Jones [4], the level of agreement between the empirical and analytical methods currently in use is limited to a specific scaling regime, i.e. short storm durations and small catchment areas. Thus, these methods are inappropriate for use with a comprehensive set of temporal and spatial scales such as at a quaternary catchment (QC) level. On the other hand, a number of these empirical (storm-centred) and analytical (correlation-based and annual maxima-centred) methods do not provide probabilistically correct areal design rainfall estimates since it is assumed that the AEP of both the point and areal design rainfall is similar [4].

In South Africa, the estimation of ARFs is limited to the storm-centred approaches of Van Wyk [16] and Wiederhold [17], while Alexander [18] also developed a geographically-centred approach based on the UK FSR methodology and observed daily rainfall data limited to the 1980s [9]. There has also been a concern in some sections of the hydrological community in South Africa that the UK FSR results may not be appropriate for South African conditions [19]. Moreover, some studies (e.g. [1], [12]) have conclusively shown that ARFs are dependent on the average AEP of rainfall.

Based on the shortcomings highlighted above, it is clearly evident that the development of ARFs appropriate to South Africa is a high-priority research area in design flood estimation. Hence, the primary objective of this study is to derive geographically-centred and probabilistically correct ARFs representative of the different rainfall-producing mechanisms at a QC level in the Modder-Riet River Basin (MRRB), South Africa as pilot study area. The focus is on the development of probabilistically correct ARFs, in other words, to establish the relationships between *T*-year areal design rainfall estimates and weighted average *T*-year design point rainfall estimates.

3 STUDY AREA

South Africa is divided into 22 primary drainage regions, which are further delineated into 148 secondary drainage regions [20]. The MRRB, as shown in Fig. 1, is situated in the C5 secondary drainage region within the primary drainage Region C and covers 34 795 km^2 [21]. The MRRB consists of two tertiary drainage regions, the Riet River (C51) and Modder River (C52) catchments, which are further sub-divided into 23 QCs.

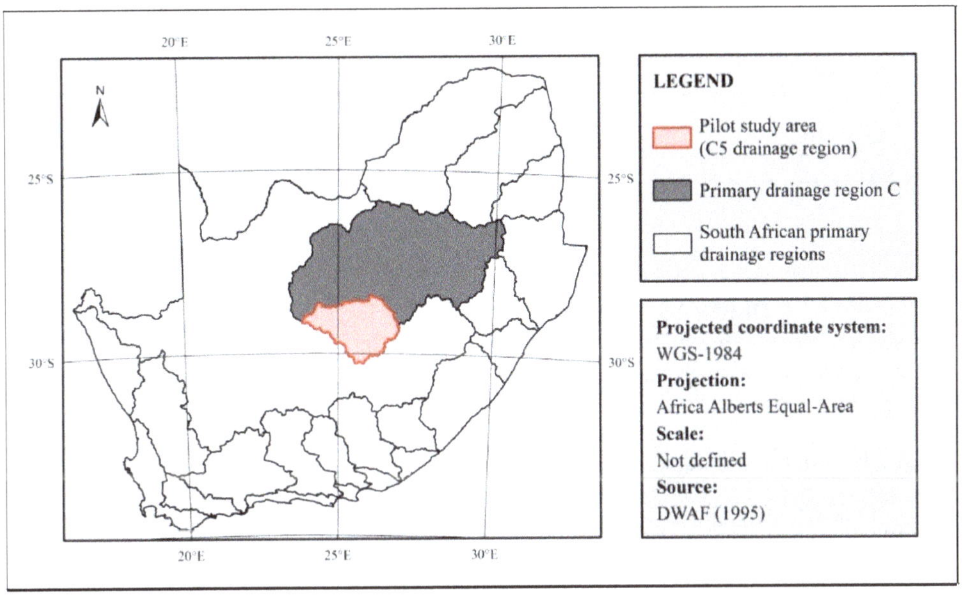

Figure 1: Location of the pilot study area [22].

The MRRB is predominantly characterised by convective rainfall associated with high rainfall intensities and thunder activity during the summer months. The Mean Annual Precipitation (MAP) is 424 mm, ranging from 275 mm in the west to 685 mm in the east [23]. The 185 South African Weather Services (SAWS) daily rainfall stations located within the boundaries of the MRRB are shown in Fig. 2. The rainfall monitoring network is in general denser in the mid-eastern parts of the study area as opposed to the north-western parts.

The overall distribution and location of the individual rainfall stations are regarded as sufficient for the purpose of this study. However, when point rainfall depths are converted to rainfall depths over an area using averaging techniques, e.g. Thiessen polygon method [24], denser rainfall monitoring networks are preferred. Consequently, the 38 neighbouring rainfall stations of the MRRB were therefore also considered in this study.

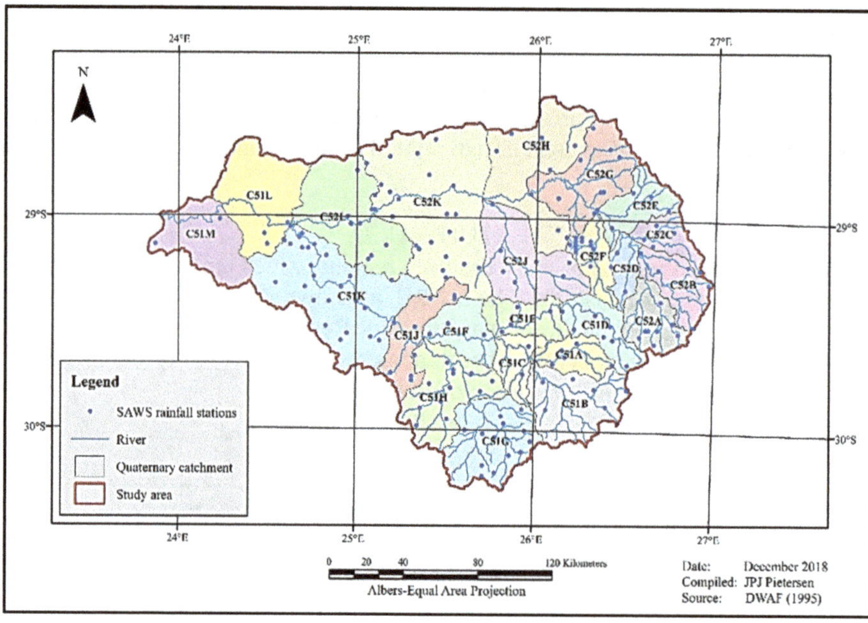

Figure 2: Location of the daily SAWS rainfall stations in the pilot study area.

4 DERIVATION OF GEOGRAPHICAL-CENTRED ARFs

This section presents the methodology adopted and estimated geographically-centred ARF results representative of the different rainfall-producing mechanisms at a QC level in the MRRB, South Africa.

4.1 Analysis of rainfall data

A daily rainfall database was established by evaluating, preparing and extracting daily rainfall data from the SAWS rainfall stations present in the MRRB as well the data from neighbouring rainfall stations. The Daily Rainfall Extraction Utility (DREU) [23] was used for the extraction and infilling of all the daily rainfall data series. Each rainfall station identified with the DREU was evaluated in terms of record length (≥ 30 years), data quality and geographical location in relation to the specific QC under consideration. The daily point rainfall annual maximum series (AMS) and areal AMS were established and extracted from the observed rainfall data for the purpose of probabilistic analyses. Infilling of daily rainfall values were necessary in some cases to obtain a minimum record length of 30 years. The observed rainfall data represents 15,791 years in total, as opposed to the 6,053 infilled years; hence, 72.3% of the total record lengths used are based on observed data.

4.2 Conversion and scaling factors

The application of conversion and/or scaling factors were considered to convert/scale the fixed 1-day rainfall to a continuous 24-hour rainfall series. Each daily observed point rainfall data series recorded at a fixed 1-day interval was converted/scaled to a continuous 24-hour rainfall series by making use of the Adamson [25] conversion and Smithers and Schulze [26] scaling factors, respectively.

The conversion factors applied to the point and areal AMS remained constant for the various durations (*D*) under consideration, e.g. 0.6 (*D* = 1-hour), 0.90 (*D* = 8-hour), 0.96 (*D* = 16-hour), 1.11 (*D* = 24-hour), 1.05 (*D* = 72-hour) and 1.02 (*D* = 168-hour). The estimated average scaling factors for each duration typically ranged from 0.56 (*D* = 1-hour), 0.92 (*D* = 8-hour), 1.05 (*D* = 16-hour), 1.19 (*D* = 24-hour), 1.38 (*D* = 72-hour) to 1.71 (*D* = 168-hour). A comparative example between the conversion and scaling factors as applicable to QC C51M is shown in Fig. 3.

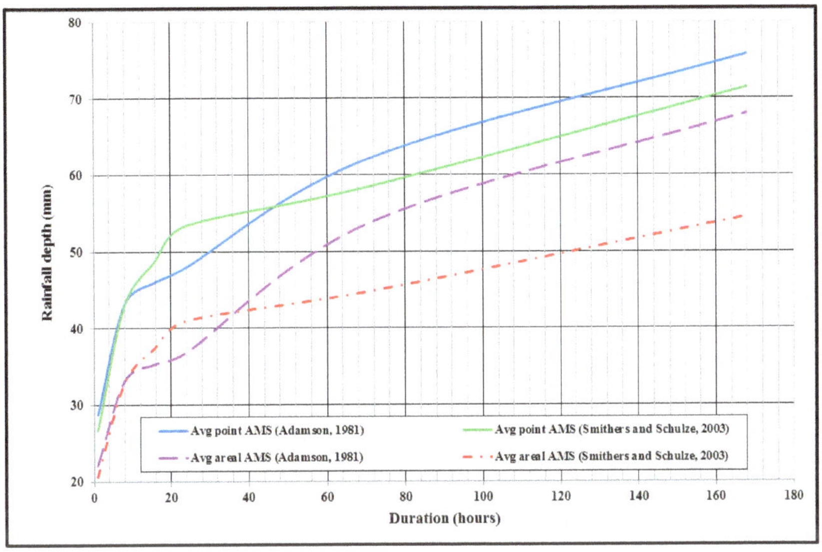

Figure 3: Comparative example of conversion and scaling factors in QC C51M.

The data presented in Fig. 3 are based on a mutual data period of 70 years applicable to seven rainfall stations located in QC C51M. From Fig. 3, it is evident that the scaling factors [27] tend to increase at a constant rate for durations > 24-hour.

4.3 Averaging of observed rainfall

Various methods proposed for the averaging of point rainfall depths over an area were considered in this study. The results obtained by [3] confirm the even spatial distribution of the rainfall stations and the relatively flat topography of the MRRB. Based on these findings, the large amount of data and computations necessary, and the preferential use of the Thiessen polygon method in various international ARF studies, e.g. Bell [28], Stewart [29] and Siriwardena and Weinmann [1], the Thiessen polygon method was selected as the most suitable method to use. The generated Thiessen polygons in the MRRB are shown in Fig. 4.

4.4 Probabilistic Analyses of Weighted AMS

The quintile estimates of point and areal rainfall are more reliable when a longer period of record is used. The probabilistic analyses of point and areal rainfall AMS were conducted separately to result in separate point and areal design rainfall frequency curves. Probabilistic analyses of the point rainfall AMS were conducted at a QC level to result in one

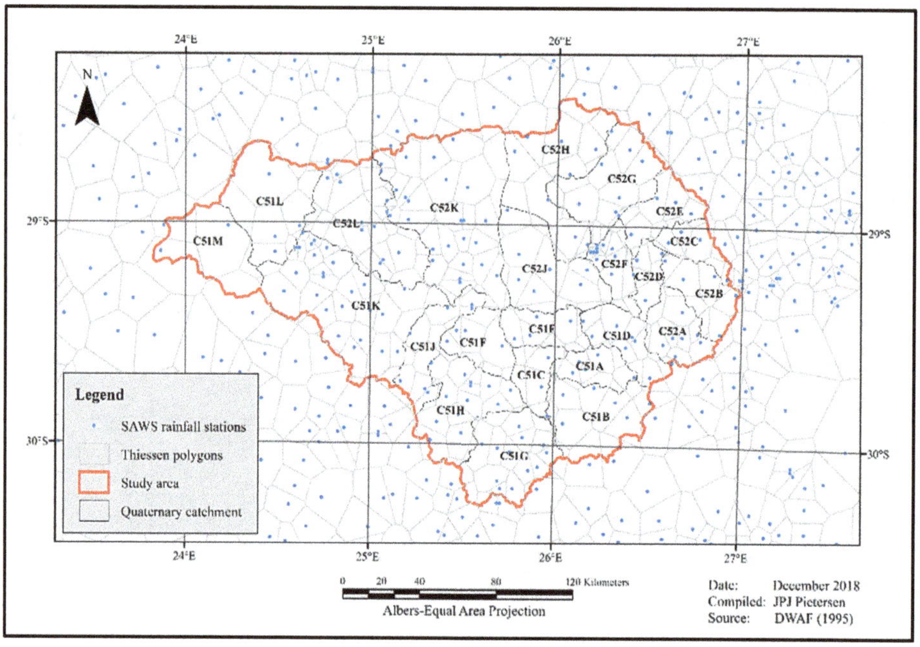

Figure 4: Layout of the Thiessen polygons in the MRRB.

representative frequency curve condensing information from all the point rainfall data series within a particular QC. Similarly, probabilistic analyses of the areal rainfall AMS (extracted and weighted at a daily time interval within a particular QC), were also conducted at a QC level to result in one representative areal frequency curve which condenses information from all the areal design rainfall data series within a particular QC.

The selection of the most suitable theoretical probability distribution was based on the statistical properties (mean, standard deviation, skewness and coefficient of variation) of each point and/or areal rainfall AMS. Typically, the Log-Normal (LN), Log-Pearson Type 3 (LP3), General Extreme Value (GEV) and Generalised Logistic (GLO) probability distributions were considered for the frequency analyses and probabilistic curve fitting. However, Smithers and Schulze [26] highlighted that the GEV probability distribution is regarded as the most suitable distribution to estimate 1-day design rainfall values in South Africa.

A typical example of the probabilistic plot results based on the ranked point AMS values in QC C51M for various durations is illustrated in Fig. 5.

4.5 Estimation of ARFs

The estimation of ARFs is based on a modified version of Bell's method [28], since the AMS of point and areal rainfall were used as opposed to the partial duration series (PDS). One set of sample ARFs was estimated for each QC with durations of 1, 8, 16, 24, 72 and 168 hours with corresponding *T*-intervals of 2, 5, 10, 20, 50, 100 and 200 years. Fig. 6 is illustrative of the typical ratio between point and areal design rainfall estimates used when individual sample ARFs need to be estimated. The variation of sample ARFs with the corresponding *T*-intervals is also shown.

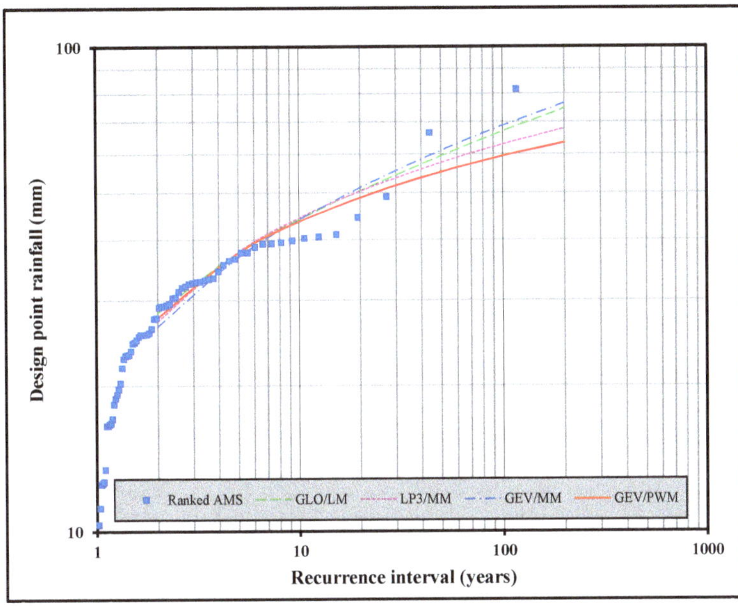

Figure 5: Example of probabilistic curve fitting to estimate design point rainfall in QC C51M.

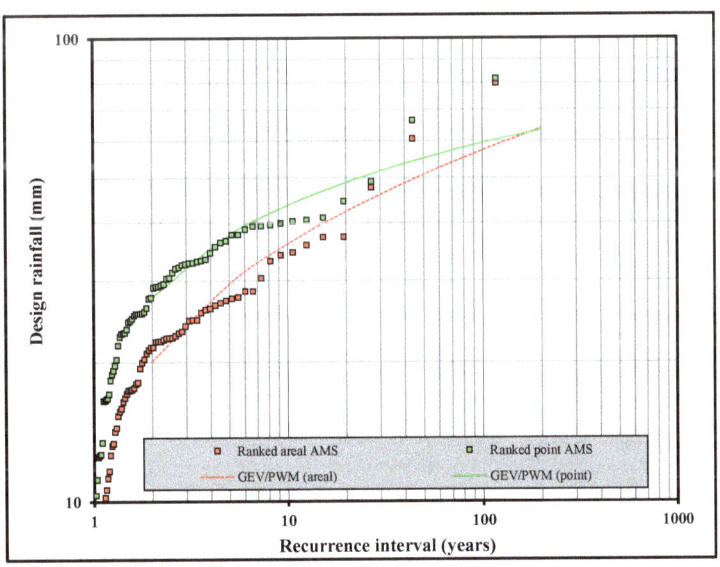

Figure 6: Example of 1-hour point and areal design rainfall in QC C51M.

It is evident from Fig. 6 that individual sample values of the geographically-centred ARFs can be expressed as the ratio between the areal catchment design rainfall and single station design point rainfall estimates for corresponding *T*-intervals using eqn (1)

$$ARF_{sample} = \frac{Areal\ design\ rainfall}{Design\ point\ rainfall}. \quad (1)$$

The ARFs estimated using eqn (1) in each QC in the MRRB were pooled together to estimate average sample ARFs for a combination of different storm durations and *T*-interval values. The latter results are listed in Table 1.

Table 1: Geographically-centred sample ARF value ranges in the MRRB.

	T **(years)**	**Storm duration (hours)**		
		< 24	**72**	**168**
Riet River Catchment (C51)	**2**	0.61–0.73	0.79–0.85	0.83–0.89
	5	0.67–0.79	0.84–0.90	0.87–0.92
	10	0.68–0.83	0.85–0.93	0.89–0.94
	20	0.68–0.87	0.86–0.95	0.91–0.97
	50	0.68–0.93	0.86–0.97	0.92–0.99
	100	0.68–1.00	0.86–1.00	0.93–1.00
	200	0.68–1.00	0.86–1.00	0.93–1.00
Modder River Catchment (C52)	**2**	0.58–0.79	0.78–0.89	0.84–0.93
	5	0.62–0.80	0.83–0.92	0.88–0.95
	10	0.65–0.82	0.86–0.94	0.91–0.97
	20	0.68–0.84	0.87–0.97	0.93–0.98
	50	0.69–0.89	0.88–0.98	0.94–0.99
	100	0.69–0.93	0.89–1.0	0.95–1.00
	200	0.69–0.97	0.89–1.0	0.96–1.00

In terms of recurrence interval, it was evident that the sample ARF values as listed in Table 1, are not constant and tend to increase with an increase in recurrence interval. The average sample ARF values ranged from 0.58 ($T = 2$-year; $D \leq 24$-hour) to 1.0 ($T \geq 100$-year; applicable to most of the durations). In some cases, the average sample ARF values deviated from the expected norm, i.e. an increase in catchment area with decreasing ARF values. Typically, larger ARF values are evident in some of the larger QCs as opposed to some of the smaller QCs.

In most of the QCs under consideration, the ratio between point and areal design rainfall for various durations, e.g. the sample ARFs, equalled unity (≈ 1) for $T \geq 100$-year. Apart from the possible presence of uniform rainfall events for the larger *T*-intervals, the areal average design rainfall depth or intensity is likely to be less than the maximum design point rainfall depths or intensities. This is confirmed by the results and also highlighted that design point rainfall estimates are only representative for a limited area.

5 CONCLUSIONS AND RECOMMENDATIONS

In this study, an enhanced methodology to express the spatial and temporal rainfall variability at a QC level by means of probabilistically correct ARFs was developed. The geographically-centred ARFs are representative of the different rainfall-producing mechanisms at a QC level in the MRRB. In addition to this, the ARF values are regarded as being probabilistically correct seeing that the relationships between *T*-year areal rainfall estimates and weighted average *T*-year point rainfall estimates for various catchment areas, storm durations, recurrence intervals and MAP values at a QC level were established.

The major findings of the study are as follows:

a) Design point rainfall estimates are only representative of a limited area and for larger areas the areal average design rainfall depth or intensity is likely to be less than the maximum design point rainfall depths or intensities.
b) ARFs vary according to predominant weather types, storm durations, climatological factors and recurrence intervals.
c) The use of a geographically-centred approach based on a modified version of Bell's method has proved to be appropriate for the study undertaken bounded within a "fixed" catchment area, namely, at a QC level.

In view of the results obtained from this study, the methodology should be expanded to other catchments in South Africa and/or internationally by incorporating the: (i) derivation of empirical ARF equations using multiple regression analysis, (ii) regionalisation of ARF equations, (iii) estimation of ARF index values to enable the transfer of hydrological information from gauged to ungauged sites, and (iv) development of a software interface to enable practitioners to apply and use the regionalised ARF equations.

REFERENCES

[1] Siriwardena, L. & Weinmann, P.E., Derivation of Areal Reduction Factors for Design Rainfalls in Victoria, Report No. 96/4. Cooperative Research Centre for Catchment Hydrology: Victoria, Australia, 1996.

[2] Alexander, W.J.R., *Flood Risk Reduction Measures: Incorporating Flood Hydrology for Southern Africa,* Department of Civil and Biosystems Engineering, University of Pretoria: Pretoria, RSA, 2001.

[3] Gericke, O.J. & Du Plessis, J.A., Evaluation of critical storm duration rainfall estimates used in flood hydrology in South Africa. *Water SA,* **37**(4), pp. 453–470, 2011. DOI: 10.4314/wsa.v37i4.4.

[4] Svensson, C. & Jones, D.A., Review of methods for deriving areal reduction factors. *Journal of Flood Risk Management*, **3**, pp. 232–245, 2010.

[5] Bengtsson, L. & Niemczynowicz, J., Areal Reduction Factors from rain movement. *Nordic Hydrology*, **17**(2), pp. 65–82, 1986.

[6] Bacchi, B. & Ranzi, R., On the derivation of the areal reduction factor of storms. *Atmospheric Research*, **42**, pp. 123–135, 1996.

[7] Sivapalan, M. & Blöschl, G., Transformation of point rainfall to areal rainfall: Intensity-duration-frequency curves. *Journal of Hydrology*, **204**(1–4), pp. 150–167, 1998.

[8] De Michéle, C., Kottegoda, N.T. & Rosso, R., The derivation of Areal Reduction Factor of storm rainfall from its scaling properties. *Water Resources Research*, **37**(12), pp. 3247–3252, 2001.

[9] NERC, Flood Studies Report. Natural Environment Research Council: London, 1975.

[10] USWB, Rainfall Intensity-Frequency Regime, The Ohio Valley, Technical Paper 29. United States Weather Bureau: Washington, DC, 1957.

[11] USWB, Rainfall Intensity-Frequency Regime, South-eastern United States, Technical Paper 29. United States Weather Bureau: Washington, DC, 1958.

[12] Omolayo, A.S., On the transposition of areal reduction factors for rainfall frequency estimation. *Journal of Hydrology,* **145**(1–2), pp. 191–205, 1993.

[13] Skaugen, T., Classification of rainfall into small- and large-scale events by statistical pattern recognition. *Journal of Hydrology,* **200**(1–4), pp. 40–57, 1997.

[14] Asquith, W.H. & Famiglietti, J.S., Rainfall areal-reduction factor estimation using annual-maxima centred approach. *Journal of Hydrology,* **230**(1, 2), pp. 55–69, 2000.
[15] Allen, R.J. & DeGaetano, A.T., Areal reduction factors for two eastern United States regions with high rain-gauge density. *Journal of Hydrologic Engineering,* **10**(4), pp. 327–335, 2005.
[16] Van Wyk, W., *Aids to the Prediction of Extreme Floods from Small Watersheds,* Department of Civil Engineering, University of the Witwatersrand: Johannesburg, 1965.
[17] Wiederhold, J.F.A., Design Storm Determination in South Africa, HRU Report No. 1/1969. Hydrological Research Unit, University of the Witwatersrand: Johannesburg, 1969.
[18] Alexander, W.J.R., Depth-Area-Duration-Frequency Properties of Storm Rainfall in South Africa, Technical Report TR103, Department of Water Affairs: Pretoria, 1980.
[19] Van der Spuy, D. & Rademeyer, P.F., *Flood Frequency Estimation Methods as Applied in the Department of Water Affairs*. DWA: Pretoria, RSA, 2018.
[20] Midgley, D.C. & Pitman, W.V., A Depth-Duration-Frequency Diagram for Point Rainfall in Southern Africa, HRU Report 2/78, University of Witwatersrand: Johannesburg, 1978.
[21] DWAF, *GIS Data: Drainage Regions of South Africa*, Department of Water Affairs and Forestry: Pretoria, 1995.
[22] Gericke, O.J. & Smithers, J.C., Review of methods used to estimate catchment response time for the purpose of peak discharge estimation. *Hydrological Sciences Journal*, **59**(11), pp. 1935–1971, 2014. DOI: 10.1080/02626667.2013.866712.
[23] Lynch, S.D., Development of a Raster Database of Annual, Monthly and Daily Rainfall for Southern Africa, WRC Report No. 1156/1/04. Water Research Commission: Pretoria, 2004.
[24] Wilson, E.M., *Engineering Hydrology,* 4th ed., Macmillan Press, London, 1990.
[25] Adamson, P.T., Southern African Storm Rainfall. Technical Report TR102. Department of Environmental Affairs and Tourism: Pretoria, 1981.
[26] Smithers, J.C. & Schulze, R.E., Development and Evaluation of Techniques for Estimating Short Duration Design Rainfall in South Africa, WRC Report No. 681/1/00. Water Research Commission: Pretoria, 2000.
[27] Smithers, J.C. & Schulze, R.E., Design Rainfall and Flood Estimation in South Africa, WRC Report 1060/01/03. Water Research Commission: Pretoria, 2003.
[28] Bell, F.C., The Areal Reduction Factor in Rainfall Frequency Estimation, Institute of Hydrology, Report No. 35. Natural Environment Research Council: UK, 1976.
[29] Stewart, E.J., Areal reduction factors for design storm construction: Joint use of rain gauge and radar data. *Association of Hydrological Sciences*, **181**, pp. 31–40, 1989.

OBJECTIVE ASSESSMENT OF LAND USE IN HYDROGRAPHICAL STUDIES

ALFREDO RAMÓN MORTE, JOSÉ TOMÁS NAVARRO CARRIÓN & ESTELA GARCÍA BOTELLA
University of Alicante, Spain

ABSTRACT

The behaviour of floods depends on two main factors: Manning's roughness coefficient and the threshold runoff, P_0. In both parameters, land cover plays a vital role in the characterisation of small streams, which traditionally are altered in the Mediterranean basin. However, in the absence of an assessment protocol that optimises the geographical information stored in the official repositories, studies in these areas tend to be subjective, depending on the personal criteria of the technicians in charge of the study. In this paper a new method is proposed for the determination of these parameters, based on the integration of the Spanish national system for the mapping of flood-prone areas (SNCZI), which was designed by the Spanish Geological and Mining Institute (IGME), and the spatial data of the Spanish land cover and use information system (SIOSE database), designed by the Spanish National Geographic Institute (IGN). The methodology generates roughness data based on objective criteria and on a threshold runoff map, which can be reviewed by the technician in charge of the study, but which is based on updated, regulated and open official data. The result is a thorough hydrological and hydraulic characterisation, which has been tested in a western Mediterranean area characterised by the complexity of land use: the foothills between Sierra Helada and Sierra de la Cortina in the Municipality of Benidorm, located in the province of Alicante, Spain.

Keywords: flood risk, SIOSE, hazard, runoff, P_0.

1 INTRODUCTION

In the study of flood-prone areas, the behaviour of water is related, among other, to two main factors: the roughness coefficient defined by Manning's n value, which demonstrates how the water will flow (hydraulic analysis), and threshold runoff, which shows how much water will flow (hydrological analysis). Threshold runoff is certainly one of the most sensitive and determining parameters in the study of flood risk.

Flood risk is defined by the amount of precipitation arriving to the superficial streams of the basin. Since the start of rain until the start of water flow, the soil absorbs water up to a limit that is determined by the threshold runoff reflected in the P_0 value, expressed in mm. Once this threshold has been reached, water infiltrates only partially, generating a surplus that will become runoff [1]. When that happens, the soil is no longer able to absorb more water and it begins to flow over the ground across poorly defined flow paths. This is a key determinant in the hydrogeomorphological analysis for calculating the rainfall accumulation needed to generate those overland flows [2], in other words, the surface runoffs.

The study of this interrelation between rainfall and the runoff generated is a common practice used for the analysis of the hazard in the Mediterranean basins [3]. The generation of surface runoff at the level of these small-sized basins clearly represents a key process in the study of flood risk [2].

Because the definition of this parameter alters significantly the flow data collected through hydrometeorological methods, it is vital that this data is accurate when assessing the surface runoff in these basins, and when calculating the flow rates that will condition their water dynamics [4]. The more realistic the definition of baseline information is, the easier it will be to achieve results in line with reality. For this reason, it is necessary to

WIT Transactions on Ecology and the Environment, Vol 234, © 2019 WIT Press
www.witpress.com, ISSN 1743-3541 (on-line)
doi:10.2495/RBM190051

ensure that the definition of the spatial variables involved in collecting this information is as realistic as possible.

Since runoff is the cause of the most difficult problems in basin management, active action is required in this part of the water cycle in order to prevent land damage caused by water [2]. An accurate characterisation is therefore vital for a deep understanding of the water cycle and behaviour of runoff when characterising a basin.

Accordingly, it was decided to choose a small-sized basin, which represented the type of basin typically found and liable to be the object of a flood risk assessment by request. This basin is located in the foothills of Sierra Helada in the Municipality of Benidorm, a province of Alicante. This is the perfect example of the type of Mediterranean basins discussed, typically ungauged and small-sized, where real flow rate data are not available and where there are high occupation and alteration rates.

2 JUSTIFICATION

When calculating the threshold runoff, it is necessary to use three spatial variables, slope, soil type and land use. The latter is the object of this detailed study. This parameter is usually calculated through the curve number method developed by SCS US Soil Conservation Service (SCS) in 1972 and adapted by Témez [5] for its use in Spain, and it was included in the reference document for its calculation in our country, "Instrucción de Carreteras 5.2" 1989 (Highway Instruction 5.2) (amended in 1990, 1998 and 2016) [2].

This document contains a table where threshold runoff was determined as a result of the combination of the three variables, and land use was the most dynamic one, which made it necessary to update information on a regular basis. From the beginning, the characterisation of land use based on this table had no choice but to resort to photo-interpretation techniques carried out by the technician in charge. For that reason, the allocation of land use was somewhat subjective, because it depended on the technician's personal criteria. The same area studied under different criteria may deliver fairly differentiated results.

This subjectivity causes questionable analysis. A method to avoid this uncertainty would be to employ official spatial data sources in the allocation of land use, such as CORINE Land Cover (CLC) and the Information System on Soil Occupation of Spain (SIOSE), which belongs to the Spanish National Plan for Land Observation (PNOT), scale 1:50.000 [6].

3 METHODOLOGY

In Spain, the background documents used in the calculation of both parameters are the *Spanish National System for the Mapping of Flood-Prone Areas (SNCZI)* [7] and the *Highway Instruction 5.2* [8]. Thanks to their update in 2016, they have taken a big step towards the avoidance of the much feared subjectivity and towards a decrease in uncertainty in the allocation of land use, and they already include tables from the CLC for the definition of Manning's roughness coefficient and for the initial threshold runoff (P_0) in mm. Although this cartographic source is the one used and recommended by competent bodies in the field, its scale of work is clearly insufficient for the basins under study. This would require more specific and detailed flood risk assessment studies, since those basins hardly ever exceed 2 km^2. For this reason, it is important to find a cartographic source to better represent this variable on a more appropriate scale.

The methodology proposed in this paper intends to go one step further and use the SIOSE database for the definition of both parameters on a suitable scale for these small-sized basins. This enables a meticulous analysis of the much sensitive parameters in flood risk assessment: threshold runoff and Manning's roughness coefficient, both which will be defined basing on updated, standardised and open official data.

In order to verify the possible variables in the result and to carry out that objective assessment of land use in hydrological studies, we have chosen an area representing the typical basins in southeast Spain, in the foothills of Sierra Helada in Benidorm, in the gently sloping watershed area, between the ravine Barranco de Barceló in the south, and the stream Rambla de Soler in the north. It is a 1.32 km^2 basin in a highly disturbed area with a high degree of anthropic occupation and which has suffered significant changes in its land use (Fig. 1).

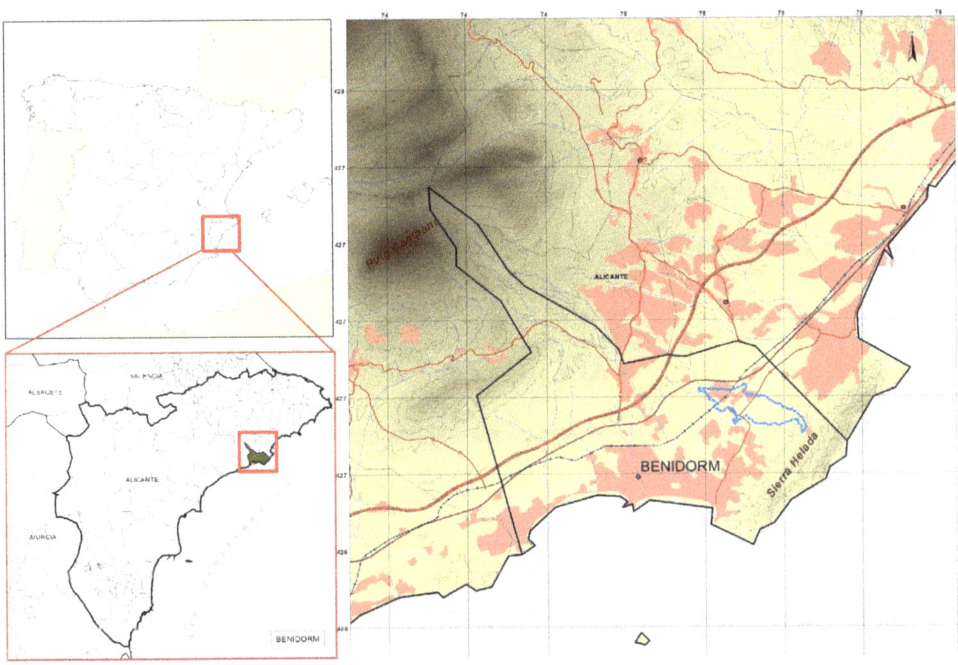

Figure 1: Location of the area under study.

3.1 Manning's roughness coefficient

The definition of Manning's roughness coefficient based on the SIOSE database is fairly simple, since land use is the only spatial variable involved. In the annexes of the SNCZI methodological guide [7], a series of tables are included where values of the Manning roughness coefficient are assigned for the land use from SIOSE as well as the use of CLC. It also includes data of common hydrological behaviour of both categorisations (CLC and SIOSE). Thus, for the assignment of Manning's value to each area using SIOSE, a calculation has been carried out combining Manning's value [9]. The result is shown in Fig. 2. This table has served as a basis to find coincidences between the hydrological behaviour of land use categorisations CLC and SIOSE.

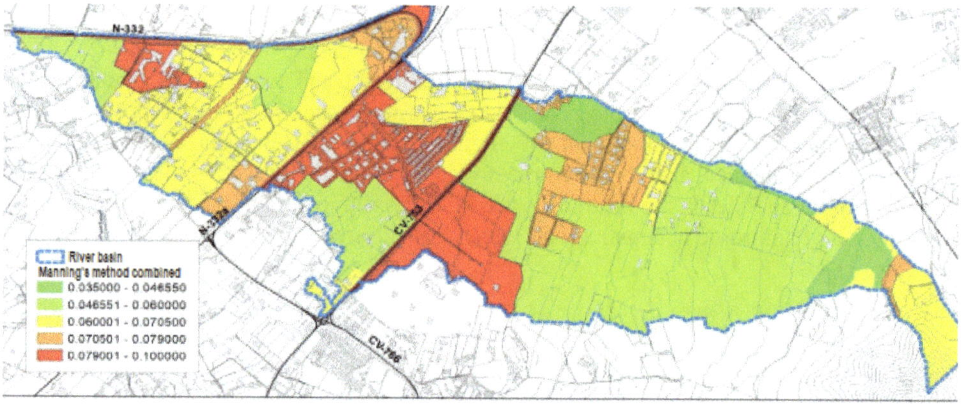

Figure 2: Manning's roughness coefficient from SIOSE.

3.2 Threshold runoff P_0 estimation

This variable is one of the most sensitive parameters in flood risk assessment studies, since a single change in its definition might lead to significant alterations in the result and, therefore, a very different flow rate, seriously affecting the flood-prone area. As mentioned in the justification, it is necessary to consider three spatial variables (slope, soil type and land use) when estimating threshold runoff P_0 estimation using SCS curve number method, and this hinders the use of SIOSE.

With the aim of providing an objective assessment of official cartographies about land use in hydrological analysis, threshold runoff P_0 has been calculated, as well as flow rates in different return periods (25, 100 and 500 years), using both, CLC and SIOSE. At the same time, the calculation of this flow rates are carried out using different hydrometeorological methods, in order to obtain a thorough characterisation that is going to allow us to draw accurate conclusions.

Finally, the same calculations are being implemented, but this time on the basis of the raster layer in Spanish threshold runoff, which can be obtained from the Ministry of Agriculture, Nature and Food Quality. Although P_0 raster layer (Fig. 3) at the Spanish level is not enough for small-sized basins (500 m pixel size), it has been included in the final comparison, just to have more baseline data.

Finally, the sources used for the calculation will be the following:

1. Calculation of the weighted threshold runoff based on the threshold runoff map of Spain.
2. Calculation of the weighted threshold runoff based on the land use layer from CLC.
3. Calculation of the weighted threshold runoff based on the land use layer from SIOSE:
   a. Manual supervised method
   b. Automated method

The sources used to characterise the other two variables involved in the calculation of threshold runoff were, on the one hand, IGN MDT05 layer for estimating the slope data using Travis' method [10] to obtain the slope threshold and, on the other hand, the Lithology layer from the former COPUT (thematic cartography of the region of Valencia in 1998) [11], for characterising soil types (Fig. 4).

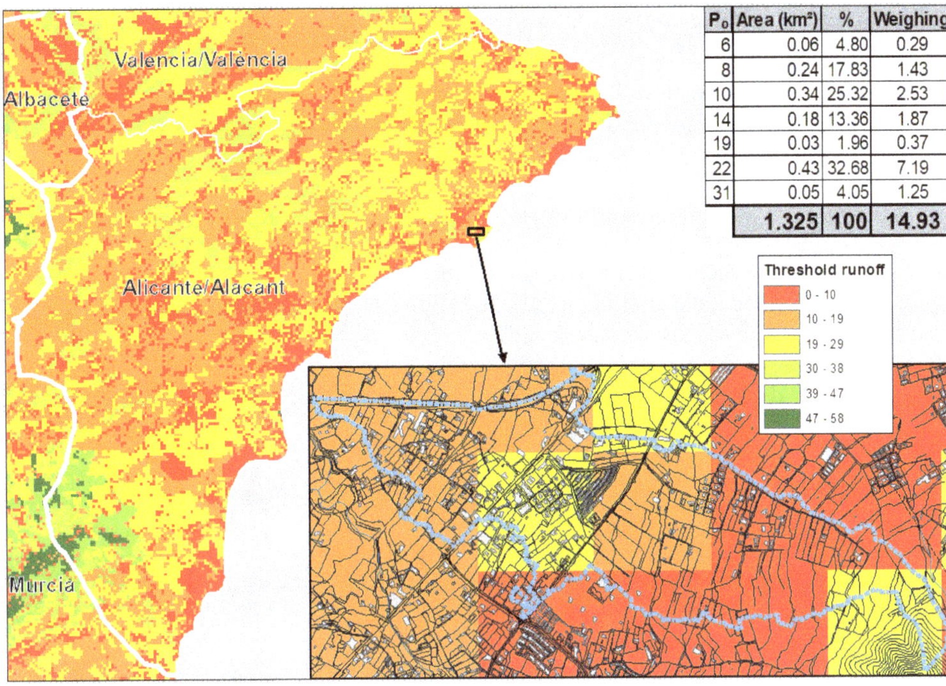

P_0	Area (km²)	%	Weighing
6	0.06	4.80	0.29
8	0.24	17.83	1.43
10	0.34	25.32	2.53
14	0.18	13.36	1.87
19	0.03	1.96	0.37
22	0.43	32.68	7.19
31	0.05	4.05	1.25
	1.325	100	14.93

Figure 3: Initial value of threshold runoff (P_0) for Spain with a pixel size of 500 m.

(a)

(b)

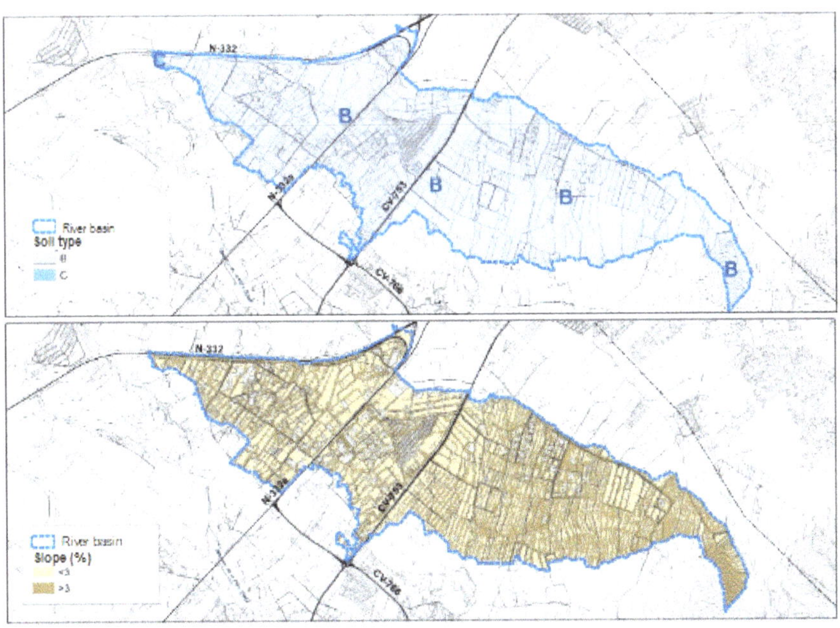

Figure 4: All other variables involved in the calculation of threshold runoff. (a) Soil type; and (b) Slopes.

The different hydrometeorological methods used to calculate flow rates are:

A. Rational method.
B. Modified Témez's method.
C. Modified Témez's method, without rainfall derating factor (where Ka = 1).
D. Highway Instruction 5.2., 2016 updated version.

3.3 Calculation of threshold runoff using CLC

In order to calculate threshold runoff applying the SCS curve method on the basis of CLC layer, the equivalence table is found in both the Highway Instruction 5.2 [8] and the SNCZI specification [7]. The results are shown in Fig. 5.

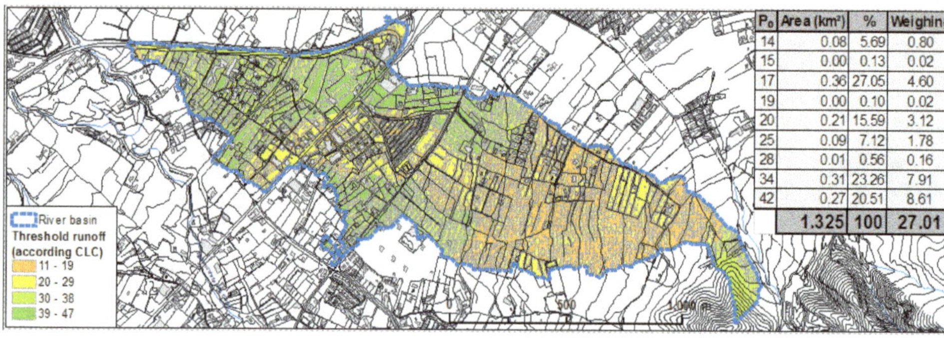

Figure 5: Threshold runoff value (P_0) calculated a on the basis of land use from CLC.

3.4 Calculation of threshold runoff based on the SIOSE database

To calculate threshold runoff on a more appropriate scale, the SIOSE database will be taken as a reference, being aware that the same area in the database may encompass several land cover observations, and that occupation rates are represented with percentages. The main feature of SIOSE is that some of its polygons do not have a unique descriptor, like CLC, but a complex codification and, thus, a richer one. Every area has its own unique identifier, named ID_POLYGON and it can be simple or compound. While the simple ones hold one descriptor, the compound areas have several land cover descriptors, they can be classified in regular covers, irregular covers and associations and their occupation rates are expressed as percentages.

This database design makes it difficult to carry out processes in which more than one spatial variable is involved, such as the calculation of P_0. For that reason, to enable the use of the rich information from SIOSE, two variants are proposed: the manual supervised method and the combined method.

3.4.1 Manual supervised method

In both cases, a unique hydrological behaviour must be allocated to a specific area from SIOSE, depending on the soil types in that area and the occupation rate in each of them. It is not a matter of setting an equivalence between CLC and SIOSE, as done in other methods [12], because it may introduce serious errors in the study of hydrological behaviour. In this case, it is required to allocate an identifier that represents the hydrological behaviour of different land uses in an specific area and, specially, their occupation rates.

Fig. 6 shows three examples of the SIOSE code. In the first one (Fig. 6(a)) the code represents a simple cover with a single identifier (Grassland – PST), which accounts for 100% of the area. In this case, allocation is simple, because the hydrological behaviour of grasslands is directly assigned to that area.

18fdd46d-b778-4ce6-af1d-8a69e387ad72	PST	2.891851	300	2.891851	100	Grassland	PST

(a)

a11ae96c-0b6f-4de4-8e2b-c456a39abce0	R(95LFCrr_05L	19.145196	222	18.187936	95	Citrus fruit trees	LFC
a11ae96c-0b6f-4de4-8e2b-c456a39abce0	R(95LFCrr_05L	19.145196	103	0.95726	5	Artificial water cover	LAA

(b)

9eb326af-1b5b-4a7f-9b29-6e3406ae4993	R(58UDS(40ZA	15.381317	102	3.568465	23.2	Wood green area	ZAU
9eb326af-1b5b-4a7f-9b29-6e3406ae4993	R(58UDS(40ZA	15.381317	101	3.122407	20.3	Edification	EDF
9eb326af-1b5b-4a7f-9b29-6e3406ae4993	R(58UDS(40ZA	15.381317	104	1.338175	8.7	Road, parking, pedestrian road	VAP
9eb326af-1b5b-4a7f-9b29-6e3406ae4993	R(58UDS(40ZA	15.381317	121	0.892116	5.8	Non built-up area	SNE
9eb326af-1b5b-4a7f-9b29-6e3406ae4993	R(58UDS(40ZA	15.381317	300	3.076263	20	Grassland	PST
9eb326af-1b5b-4a7f-9b29-6e3406ae4993	R(58UDS(40ZA	15.381317	222	1.538132	10	Citrus fruit trees	LFC
9eb326af-1b5b-4a7f-9b29-6e3406ae4993	R(58UDS(40ZA	15.381317	212	1.076692	7	Arable crops (not rice)	CHL
9eb326af-1b5b-4a7f-9b29-6e3406ae4993	R(58UDS(40ZA	15.381317	111	0.769066	5	Other structures	OCT

(c)

Figure 6: Table of attributes from SIOSE. (a) Example of an area from SIOSE with a land use of 100%; (b) Example of compound cover from SIOSE with a land use distribution of 90%; and (c) Example of an area with very similar occupation rates and with a significantly different hydrological behaviour.

In the second example (Fig. 6(b)) another usual combination is depicted, since it is a common land cover pattern in certain areas, increasing their occupation rate. In this case, the hydrological behaviour is determined by the most representative land cover in terms of its occupation rate, fruit trees in this case (Citrus – LFC), with a distribution of 95%.

The third example (Fig. 6(c)) poses a classification problem, since there are compound covers with a wide variety of descriptors which entail highly differentiated hydrological behaviours, but with similar occupation rates. In this example, there are three differentiated hydrological behaviours (low thresholds of around 1 mm, intermediate thresholds of 35 mm and thresholds varying from 20 to 25 mm). Each of these three groups represents a 33% of the occupation rate in the area. Therefore, the allocated hydrological behaviour must be in between both extremes and towards the safest threshold in terms of flood risk assessment.

This manual assignment (Fig. 7) still has certain degree of uncertainty. As a solution, the combined method is proposed below.

3.4.2 Combined method

In order to prevent subjectivity in the manual method when assigning values, a second method named "combined method" (Table 1) is proposed, inspired by the one used for the assignment of Manning's roughness value.

In this method, the calculation is done on the basis that all information likeable to prevent this subjectivity is included in the table, that is, every land cover descriptor for each area along with its corresponding hydrological behaviour and occupation rate.

Intersecting slope an soil type layers with SIOSE land cover data poses a main drawback: the exact location of each member within a SIOSE compound cover is unknown. Hence we propose a tradeoff that consists of weighing each cover according to the occupation rates of each slope category and soil type within a SIOSE area.

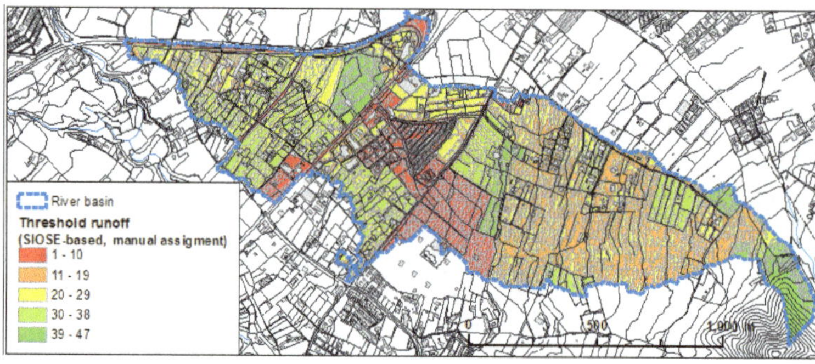

P_0	Area (km²)	%	Weighing
1	0.24	18.10	0.18
8	0.00	0.04	0.00
12	0.00	0.02	0.00
14	0.36	26.99	3.78
15	0.00	0.05	0.01
19	0.00	0.09	0.02
20	0.02	1.20	0.24
23	0.00	0.34	0.08
25	0.24	18.01	4.50
28	0.17	12.79	3.58
34	0.18	13.80	4.69
35	0.00	0.17	0.06
42	0.01	1.08	0.45
47	0.10	7.33	3.44
	1.325	**100**	**21.04**

Figure 7: Threshold runoff value (P_0) based on land use data from SIOSE, manual supervised method.

Table 1: Threshold runoff value (P_0) based on land use data from SIOSE, combined method.

P_0	Area (km²)	%	Weighing
0	0.01	0.39	0.00
1	0.34	25.65	0.26
8	0.00	0.01	0.00
12	0.00	0.00	0.00
14	0.16	12.21	1.71
15	0.00	0.05	0.01
19	0.00	0.09	0.02
20	0.02	1.23	0.25
23	0.04	2.67	0.61
25	0.13	9.56	2.39
28	0.16	12.23	3.42
31	0.00	0.02	0.01
34	0.18	13.40	4.56
35	0.03	2.20	0.77
42	0.01	1.12	0.47
47	0.25	19.18	9.01
	1.325	**100**	**23.48**

4 RESULTS

As mentioned in the methodology section, to perform a thorough assessment of results, it has been decided to also include the results from the calculation based on IGME map for Spain, as well as CLC land use data. At the same time, the flow rate details for different return periods have been calculated using the four hydrometeorological methods that were previously mentioned (method A, B, C and D). The final result from the different calculation sources and methods is given in Table 2.

As seen in Table 2, the lowest threshold is the result from the calculation for Spain (Source 1), which is 14.93 mm. This calculation has low accuracy, because the pixel size is 500 m, clearly not enough for small-sized basins. Flow rates are much higher, twice the value of the other calculations. Therefore, even though this is not the most appropriate scale, the flow rate result is higher, and we keep in the safe side.

Table 2: Table of results.

	Return period									P_0
	25 years			100 years			500 years			
	Methods			Methods			Methods			
	A	B and D	C	A	B and D	C	A	B and D	C	
Source 1	23.98	19.74	20.1	45.2	37.24	37.8	78.3	64.69	65.6	**14.93**
Source 2	9.16	7.47	7.67	22.2	18.24	18.6	44.9	36.92	37.6	**27.01**
Source 3a	14.88	12.2	12.5	31.3	25.76	26.2	58.5	48.23	49	**21.04**
Source 3b	12.27	10.04	10.3	27.2	22.36	22.8	52.4	43.16	43.9	**23.48**

The outcome from Source 2 (CLC) yields the highest threshold value (27.01 mm), and for that reason it will provide lower flow rates. Although this is the method suggested by the competent government bodies in this field, baseline data are not accurate enough, and, when threshold runoff value increases, which means more water infiltrating the soil before runoff generation, flow rates decrease, and thus yield the lowest results.

The next values correspond to the analysis performed with the Source 3 (SIOSE), which guarantee an adequate scale, a higher thematic resolution, updates every 3 years, open and standardised official data. On the other hand, SIOSE's descriptive data model requires additional processing since it hampers one-to-one relations with the other two variables (slope and land type) involved in the calculation of P_0. The results from the calculation with this source for the manual method (3a) and the combined method (3b) are 21.04 mm and 23.48 mm respectively. Even though, the manual method brings somewhat subjectivity, since the assignment criteria depends on the technician in charge, this assignment remarkably tends to remain always on the safe side. Finally, although one must bear in mind that in the combined method, the distribution of the other two variables (land type and slope) will affect to every land cover represented in an area, this method remains the one providing the highest level of objectivity to the calculation.

5 DISCUSSION

As for the methods proposed in this paper, the manual method implies subjectivity, since it is based on a supervised assignment, which depends on the technician's skills, experience and expertise. A correct implementation of this method requires previously assigning hydrological behaviour to a training set of areas so that subsequent extrapolation is made based on well-defined criteria. This way, assignment discrepancies between similar areas can be narrowed. Indeed, a key point in manual assignment is to design a protocol on these criteria to be applied in similar areas and, thus, prevent subjective assessment by extrapolating this estimation to the characterisation of areas that meet the same conditions.

Nevertheless, the manual assignment of values remains highly subjective, even if it is done cautiously, valuing greater certainty and trying to continually learn in the assignment in order to extrapolate the information to similar areas. Furthermore, this method implies a simplification of the information collected from SIOSE areas, which means discarding SIOSE's thematic resolution, and even worse, producing a reclassification of the values which tends to replicate the structure of CLC.

On the other hand, the combined method is by far the most objective one. It preserves information richness and encourages the use of updated, open, standardised official data on an appropriate working scale. The result of P_0 calculation with this method is 23.48 mm, in between the values obtained with CLC (27.01 mm) source used in SNCZI and the 21.04 mm from the manual method calculation. This result is the most logical one, since its value is between the one obtained from a cartography with an insufficient scale and one obtained from the manual method, which tends to stay on the safe side.

Once the methodology has been tested, the next step will be the automation of the process. However, before proceeding, the application of each methodology should be extended to a greater number of basins. In this way, more reference values will be available, which will allow a thorough understanding of the characterisation of hydraulic behaviour based on SIOSE land use data. In the end, working with detailed cartography like SIOSE (1:25.000 until 2018 and 1:10.000 since 2018), will provide the accuracy that the model lacks and a richer results in the flood risk assessment.

However, it has to be borne in mind that the accuracy obtained in the calculation of threshold runoff using the SIOSE database is not useful at all if the other spatial variables are not estimated with the same accuracy. The *soil type* variable, in which there are only four types defined by hydrological group, is based on a 1:50.000 scale cartography, which causes serious inaccuracies in the baseline data of the areas under study. An official cartography with an appropriate scale is therefore essential for the representation of the variable. In general, the use of official cartography as a reference is vital for such studies and for the definition of parameters, since it smooths the work, guarantees quality, and prevents human mistakes and any kind of subjectivity in the calculation.

Finally, a thorough understanding of the hydrological behaviour of land use is vital for the integral management of basins, and it is not just about the already existing land uses, but also about the newly proposed ones. A detailed analysis of the use of a basin would enable a coherent allocation of new areas and would make it easy to know which new land uses can be allowed without increasing flood risk. In this sense, the analysis of threshold runoff in a specific drainage basin should be a key tool to correct the errors of the past or, at least, to avoid remaking them, with the ultimate goal of achieving a responsible land-use planning.

ACKNOWLEDGEMENTS

The authors are grateful to the SIOSE-INNOVA Project (CSO2016-79420-R AEI/FEDER UE) and Interuniversity Institute of Geography, University of Alicante.

REFERENCES

[1] Bescos Atin, A., Camarasa Belmonte, A.M. & Al, E., Elaboración de cartografía de zonas inundables. Aplicación al Llano de Inundación del Río Arga (Navarra). *Ser. Geográfica*, **9**, pp. 219–236, 2000.

[2] Camarasa Belmonte, A.M., García López, M.J. & Pascual, J.A., Análisis mediante SIG de los parámetros de producción de escorrentía. *El acceso a la inf. Espac. Y las nuevas tecnol. Geográficas*, pp. 49–63, 2006.

[3] Almánzar, R. & Pascual, J.A., Análisis de sucesos productores de caudal en pequeñas cuencas mediterráneas como apoyo a la gestión integrada y sostenible del agua. *VII Congr. Ibérico sobre Gestión y Planif. del Agua Ríos Ibéricos +10*, pp. 1–6, 2011.

[4] López García, M.J., Camarasa Belmonte, A.M. & Mateu Bellés, J.F., Cambios en los usos del suelo y producción de escorrentía en ramblas mediterráneas: Carraixet y Poyo (1956–1998). *Bol. la Asoc. Geogr. Esp.*, **44**, pp. 69–94, 2007.

[5] Témez, J.R., Cálculo hidrometeorológico de caudales máximos en pequeñas cuencas naturales, MOPU Dirección General de Carreteras, 1978.

[6] Fomento, Geoportal SIOSE, 2019. www.siose.es/.

[7] Sánchez Martínez, F.J., Lastra Fernández, J., Andrés Moreno, C., García-Olivo González, M., Marcuello Olona, C. & Jiménez Álvarez, A., *Guía metodológica para el desarrollo del Sistema Nacional de Cartografía de Zonas Inundables*, Spanish Ministry, 2011.

[8] Ministry of Public Works and Urban Planning. Spanish Government, Spanish normative: Orden FOM/298/2016, de 15 de febrero, por la que se aprueba la norma 5.2 – IC drenaje superficial de la Instrucción de Carreteras. *Boletín of. del Estado*, **60**, March 10, pp. 18882–19023, 2016.

[9] González, J., García, O., Crespo, A., Gómez, M. & Cea, L., Predicción de inundaciones originadas por precipitaciones extremas mediante el módulo hidrológico de Iber. *V. Jornadas Ing. del Agua*, pp. 1–15, 2017.

[10] Castillejo, I., Fernández, P., Garcia-Ferrer, A. & Sánchez, M., Estudio comparativo de los distintos métodos de cálculo de pendientes derivados de modelos digitales de elevación, **1**, p. 13, 2004.

[11] Martínez J. & Balaguer, J., *Litología, aprovechamiento de rocas industriales y riesgo de deslizamiento en la Comunidad Valenciana*, Publicaciones Divulg. Técnica, p. 66, 1998.

[12] Cantarino Martí, I., Elaboración y validación de un modelo jerárquico derivado de SIOSE. *Rev. Teledetec.*, **39**, pp. 5–21, 2013.

DEVELOPMENT AND ASSESSMENT OF A SINGLE-EVENT DESIGN FLOOD ESTIMATION TOOL: CASE STUDY IN FOUR CLIMATOLOGICAL REGIONS OF SOUTH AFRICA

OCKERT J. GERICKE
Unit for Sustainable Water and Environment, Department of Civil Engineering,
Central University of Technology, South Africa

ABSTRACT
The estimation of design floods characterised by a specific magnitude-frequency relationship at a particular site is necessary for the planning, design and operation of hydraulic structures. Both the occurrence and the frequency of floods, along with the uncertainty involved in the estimation thereof, contribute to the practitioners' dilemma to make a single, justifiable decision based on the results obtained from the various single-event design flood estimation methods available internationally. This paper presents the further development and assessment of a Design Flood Estimation Tool (DFET) in four climatologically different regions of South Africa. The DFET was developed and programmed using Microsoft Visual Basic for Applications and the functionality and application thereof were assessed by comparing the DFET-based probabilistic, deterministic and empirical design flood estimates in 12 gauged catchments. The refined and improved DFET proved to be an easy to use software tool for the rapid estimation and assessment of alternative design flood estimation methods for at-site applications in both gauged and ungauged catchments. In general, the empirical methods resulted in more acceptable design flood peaks than the deterministic methods when compared to the at-site probabilistic flood peaks. Furthermore, the use of more complex deterministic methods, e.g. Synthetic Unit Hydrograph (SUH) method, as opposed to the more simplified deterministic methods, e.g. Rational method (RM) and Soil Conservation Services (SCS) method, does not necessarily guarantee and result in more acceptable peak discharge estimates. The latter also emphasises that no single design flood estimation method could be regarded as superior to any other method when the wide range of flood magnitude-frequency problems encountered in practice, needs to be solved. Practitioners still have to apply their own experience, knowledge and skills to these particular problems until the current National Flood Study Programme (NFSP) narrows the gap between flood research and practice in South Africa.
Keywords: design flood, design rainfall, estimation, flood magnitude-frequency, flood modelling.

1 INTRODUCTION

The estimation of design floods characterised by a specific magnitude-frequency relationship at a particular site is necessary for the planning, design and operation of hydraulic structures [1]. Both the occurrence and frequency of floods, along with the uncertainty involved in the estimation thereof, contribute to the practitioners' dilemma to make a single, justifiable decision based on the results obtained from the various single-event design flood estimation methods available internationally [2].

In South Africa, three basic approaches to design flood estimation are available, e.g. probabilistic, deterministic and empirical methods [3], [4]. Apart from the probabilistic methods, there is no guarantee that any of the deterministic and/or empirical methods would result in more acceptable flood magnitude-frequency relationships. Hence, in order to assess the uncertainty in using the latter design flood estimation methods, all approaches should be included in any specific design situation and not be limited to using only the more simplified methods, e.g. the +160-year old Rational method [2]. In order to overcome some of the

www.witpress.com, ISSN 1743-3541 (on-line)
doi:10.2495/RBM190061

inherent limitations of the current available methods and to enhance the practitioners' decision-making process, in conjunction with the overall objective of the National Flood Studies Programme (NFSP) in South Africa [5] to update and enhance design flood estimation methodologies, the Utility Programs for Drainage (UPD) software [6], [7] and Design Flood Estimation Tool (DFET) [2] were developed and are frequently updated.

The overall objective of this paper is to present the further development and assessment of the DFET in 12 gauged catchments located in four climatologically different regions of South Africa. The specific objectives are to: (i) refine and improve the DFET to ensure that it is compatible with other relevant software interfaces and contains the latest design rainfall information, and catchment parameter and design flood estimation methodologies, and (ii) demonstrate the use and functionality of the DFET by assessing the performance of the various design flood estimation methods in different catchment size ranges, e.g. $A < 150$ km², $150 \leq A \leq 1{,}000$ km² and $1{,}000 \leq A \leq 10{,}000$ km².

A summary of the study area is included in the next section. Thereafter, the methodologies involved in meeting the objectives are detailed, followed by the results and the discussion thereof. In the conclusion, some concluding remarks are made in terms of the major findings.

2 STUDY AREA

South Africa is located on the most southern tip of Africa and demarcated into 22 primary drainage regions (A to X) as shown in Fig. 1. These primary drainage regions are further delineated into 148 secondary drainage regions, i.e. A1, A2, to X4 [8]. The 12 gauged catchments in this study are located in 10 of these secondary drainage regions which are located in four distinctive climatological regions of South Africa, i.e. the Northern Region (NR), Central Region (CR), Southern Winter Coastal Region (SWCR), and Eastern Summer Coastal Region (ESCR) [9].

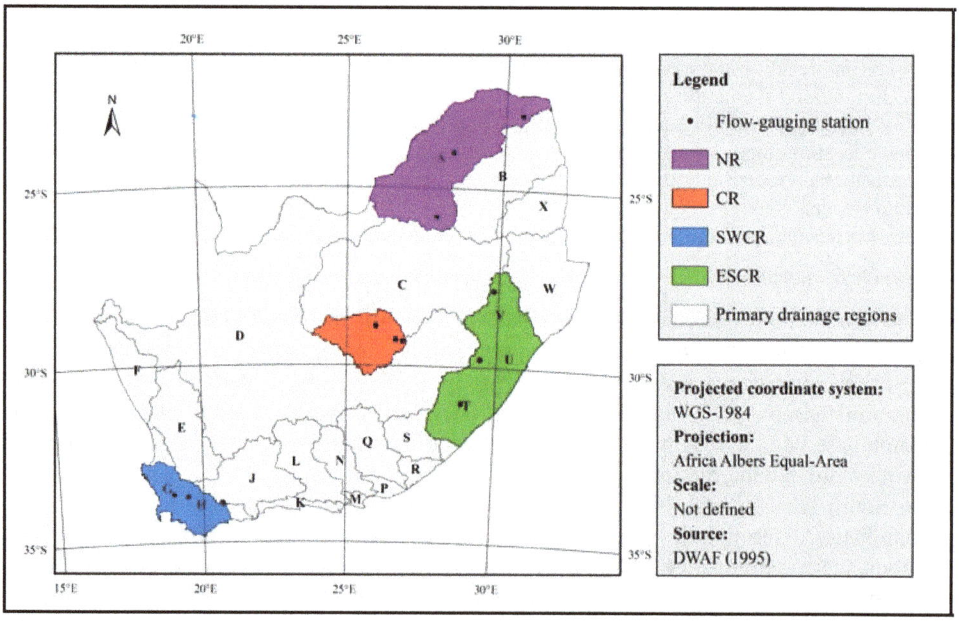

Figure 1: Location of 12 flow-gauging stations within the four climatological regions. *(Source: [10].)*

The four climatological regions are representative of the broad variations in climate (e.g. Mean Annual Precipitation (MAP), rainfall type, distribution and rainfall seasonality), catchment geomorphology, channel geomorphology, geographical location, and altitude above mean sea level found in South Africa [9]. The catchment areas range between 28 km² and 6 331 km² and are regarded as "gauged", since Department of Water and Sanitation (DWS) flow-gauging stations are located at the outlet of each catchment. Table 1 contains a summary of the main geomorphological catchment properties, e.g. MAP, catchment area (A), hydraulic length (L_H), centroid distance (L_C), average catchment slope (S) and main river slope (S_{CH}).

Table 1: Main geomorphological properties of the 12 catchments in the four climatological regions. *(Source: [9].)*

Region	Catchment descriptor	Catchment characteristics					
		MAP (mm)	A (km²)	L_H (km)	L_C (km)	S (%)	S_{CH} (%)
NR	A2H012	690	2 555	57.4	22.1	5.30	0.69
	A5H004	623	636	68.4	37.4	8.73	0.71
	A9H003	967	61	16.3	10.7	15.87	1.16
CR	C5R003	549	937	53.8	31.1	5.04	0.27
	C5R004	518	6 331	186.7	106.4	4.19	0.13
	C5R005	660	116	16.2	7.9	5.50	0.90
SWCR	G1H007	899	724	55.5	29.0	26.21	0.46
	H4H006	450	2 878	109.9	26.9	29.21	0.47
	H7H004	566	28	15.7	7.5	31.28	4.54
ESCR	T3H005	877	2 565	160.2	86.7	21.42	0.45
	T5H004	1 060	537	67.4	23.9	22.66	0.77
	V3H007	898	129	24.9	16.9	15.73	0.93

3 METHODOLOGY AND RESULTS

This section presents the methodology adopted to further develop and refine the DFET and the assessment and discussion of the DFET-based design flood estimation results at a catchment level in the four climatologically different regions of South Africa.

3.1 Development of DFET

The original DFET [2] was developed and automated using Microsoft Visual Basic for Applications (VBA). The integral part of automation is dependent on the development of the VBA project comprising of a set of modules which contains macros consisting each of a set of declarations followed by procedures or methods acting on toolbar controls. The latter controls are used to receive user input, display output and trigger event procedures.

Both interactive (responsive to user actions) and static (accessible only through code) controls are incorporated in the refined version of the DFET (currently Version 1.3). Both the information processing and application phases of the DFET are now also characterised by a full graphical interface, enabling the printing/plotting of worksheets and graphs, while an improved selection of Geographical Information Systems (GIS)-based maps for easy reference is also available. The DFET was further refined by updating the design rainfall information and methods used to estimate design rainfall. Typically, direct imports from the Regional Linear Moment Algorithm and Scale Invariance (RLMA&SI) design rainfall software [11], [12] are now possible. The catchment parameter estimations functionalities and data imports from GIS-based software were also updated to ensure an improved characterisation of various catchment parameters. For example, extensive GIS-based longitudinal profiles of main watercourses/rivers obtained from Digital Elevation Models (DEMs) in the ArcGIS™ environment could now be imported into the DFET to enable the rapid estimation of average main watercourse/river slopes using the methods as proposed by [4] and [13], respectively.

The HOME page of the DFET is shown in Fig. 2 and enables the viewing and/or editing of the relevant databases, design tables, design flood estimation methods, GIS-based maps and graphical plots contained in the various worksheets. It also serves as the primary worksheet with click buttons which activate macros to direct or redirect the user to any required worksheet.

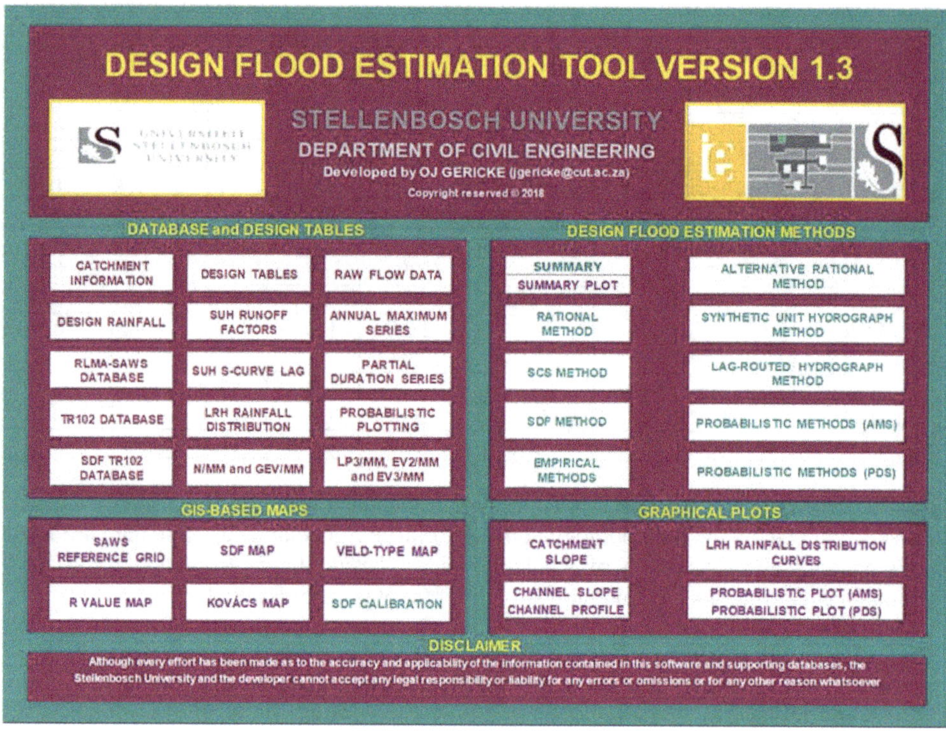

Figure 2: DFET HOME page.

3.2 Estimation of at-site probabilistic design floods

At-site probabilistic flood frequency analyses of the annual maximum series (AMS) were conducted using the DFET at the 12 flow-gauging stations to summarise the observed flood peaks, estimate parameters and select appropriate theoretical probability distributions. The observed flood peaks were summarised by ranking the AMS in a descending order of magnitude and the Cunnane plotting position [7] was used to assign return period (T) values to the plotted values. The Method of Moments (MM) and Linear Moments (LM) were used to estimate parameters to ultimately enable the fitting of theoretical probability distributions to the AMS values.

The average AMS record length of the 12 catchments is 60 years. The statistical properties of each AMS dataset also confirmed the asymmetrical nature thereof, i.e. a high degree of variability and skewness. Consequently, the General Extreme Value (GEV) and Log-Pearson Type 3 (LP3) probability distributions were regarded as the most suitable distributions in 42% and 33% of all the catchments, respectively. The Generalised Logistic (GLO) probability distribution proved to be the most appropriate distribution in the remainder 25% of the catchments. The DFET-based probabilistic plots at a catchment level in the four climatological regions are shown in Fig. 3.

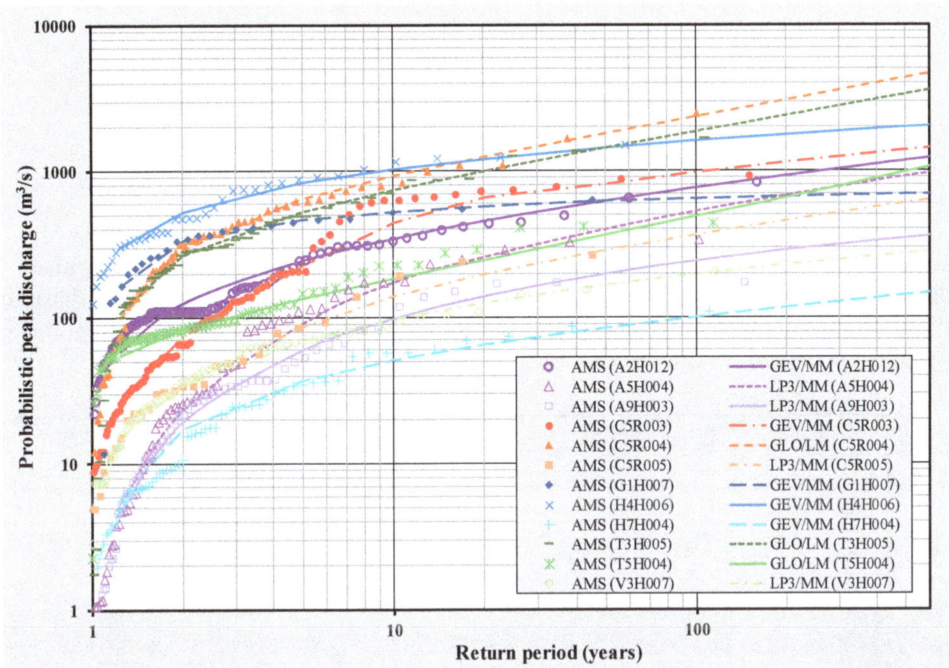

Figure 3: Probabilistic plots ($1 \leq T \leq 1{,}000$-year) based on the ranked AMS and Cunnane plotting position in the four climatological regions.

3.3 Estimation and assessment of deterministic and empirical design floods

The DFET was used to process all the catchment parameter and design rainfall information to serve as input to the various deterministic and empirical design flood estimation methods. All other design flood calculation procedures are fully automated and the standard procedure and techniques associated with each method were used by default, while taking cognisance of the assumptions, areal limitations and intended application of each method. Both single-event deterministic (e.g. Rational method (RM), Alternative Rational method (ARM), Soil Conservation Services (SCS), Standard Design Flood (SDF), Synthetic Unit Hydrograph (SUH) and Lag-Routed Hydrograph (LRH)) and empirical (e.g. Midgley and Pitman (MIPI), Catchment Parameter (CAPA) and Regional Maximum Flood (RMF)) design flood estimation methods are included in the DFET.

The DFET-based design flood estimation results (Q_D) based on both the deterministic and empirical methods compared to the at-site flood frequency analyses (Q_P) for return periods ranging from two to 200 years in each catchment were expressed as peak discharge ratios using eqn (1). The average peak discharge ratios (eqn. (1)) at a catchment level are summarised and shown in Fig. 4. In addition, the Root-Mean-Square-Error (*RMSE*) estimated using eqn (2) and shown in Fig. 5 was also considered to ensure that the accumulated over- and/or underestimations are accounted for, i.e. to highlight the actual size of errors produced by each method, with the objective function to minimise the *RMSE* to zero.

$$Q_{T-ratio} = \left(\frac{Q_D}{Q_P} \right) - 1, \tag{1}$$

$$RMSE = \sqrt{\frac{\sum_{i=1}^{N} \left(Q_{Pi} - Q_{Di} \right)^2}{N}}, \tag{2}$$

where $Q_{T\text{-}ratio}$ is the peak discharge ratio (positive = overestimation and negative = underestimation), *RMSE* is the Root-Mean-Square-Error, Q_D is the design peak discharge (m^3/s) estimated using either a deterministic or empirical method, and Q_P is the at-site probabilistic design peak discharge (m^3/s).

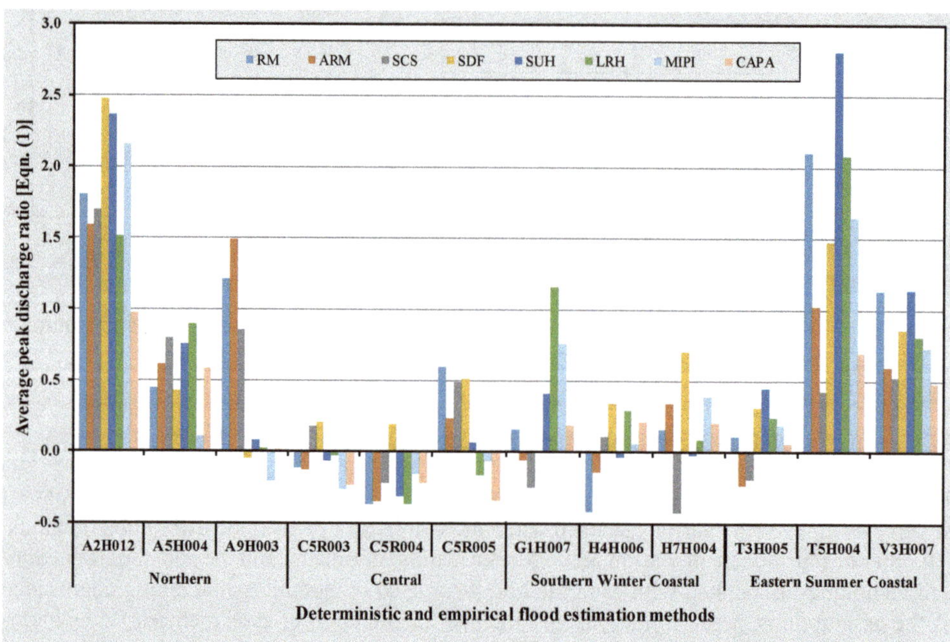

Figure 4: Average peak discharge ratios (eqn (1)) at a catchment level in the four climatological regions.

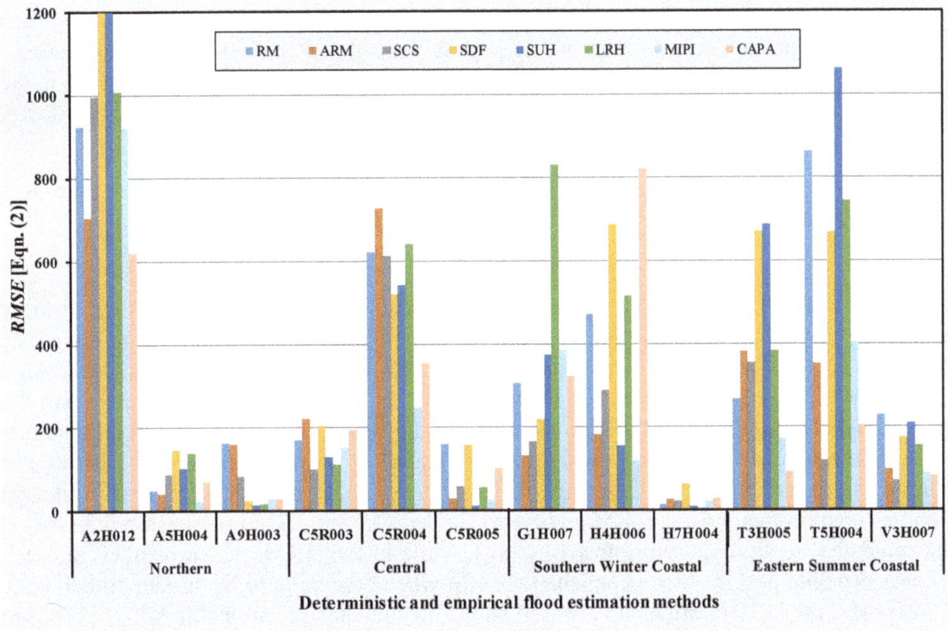

Figure 5: RMSE statistics (eqn (2)) based on the comparison between the probabilistic, deterministic and empirical peak discharges at a catchment level in the four climatological regions.

The results contained in Figs 4 and 5 are indicative of several trends associated with specific catchments and return periods, which are highlighted below:

- **Northern Region:** On average, the deterministic and empirical flood peaks (Q_D) exceeded the at-site probabilistic flood peaks (Q_P) in both catchments A2H012 and A5H004, respectively. In catchment A9H003, all the methods also exceeded the Q_P values, except for the SDF and MIPI methods. On average, the empirical methods, especially the CAPA method, demonstrated the best results, e.g. 1%–97% overestimation, $r^2 \approx 0.99$, and $30 \leq RMSE \leq 623$.
- **Central Region:** In this region, the results varied from overestimations to underestimations $< 50\%$. The SDF flood peaks constantly exceeded the at-site probabilistic flood peaks in all the catchments, followed by the SCS method in two catchments. In catchment C5R005, all the deterministic methods, except the LRH method, exceeded the Q_P values. In all other cases, the at-site probabilistic flood peaks were underestimated. On average, the MIPI method demonstrated the best results, e.g. -6%–-26% underestimation, $0.93 \leq r^2 \leq 0.99$, and $26 \leq RMSE \leq 248$.
- **Southern Winter Coastal Region:** On average, the deterministic flood peaks based on the ARM (catchments G1H007 and H4H006), SCS (catchments G1H007 and H7H004), RM and SUH (catchment H4H006) underestimated the at-site probabilistic flood peaks, while all the other methods resulted in overestimations $\leq 75\%$. On average, the ARM demonstrated the best results, e.g. -6%–34% underestimation/overestimation, $0.80 \leq r^2 \leq 0.83$, and $26 \leq RMSE \leq 183$.

- **Eastern Summer Coastal Region:** On average, the deterministic and empirical flood peaks exceeded the at-site probabilistic flood peaks in all the catchments, except for the ARM and SCS methods in catchment T3H005. On average, as in the case of the Northern Region, the CAPA method demonstrated the best results, e.g. 6%–70% overestimation, $r^2 \approx 0.87$, and $81 \leq RMSE \leq 203$.

Above-listed results highlight a number of important issues that need to be discussed further in the next section.

4 DISCUSSION

In most of the catchments, the empirical methods (e.g. MIPI and CAPA) resulted in more acceptable design flood peaks than the deterministic methods when compared to the at-site probabilistic flood peaks. The fact that the empirical methods are less reliant on design rainfall information, site-specific runoff coefficients and catchment response time than the deterministic methods, contributed towards this trend. In essence, this trend highlights that subjectivity could be introduced when design rainfall is estimated using different depth-duration-frequency (DDF) relationships in conjunction with specific deterministic design flood estimation methods, e.g. RM and ARM. In the RM, the DDF relationship is based on the 'outdated' methodology proposed by [14], while in the ARM, which resulted in much more acceptable design flood peaks than the RM when compared to the at-site probabilistic flood peaks, the DDF relationship is based on the updated and recent methodology proposed by [11], [12]. Furthermore, the selection of site-specific runoff coefficients or curve numbers, i.e. coefficients representative of the percentage of effective rainfall which is converted into direct runoff, not only introduces more subjectivity and associated uncertainty, but wrongfully also "ignore" the antecedent moisture conditions by being constant for a specific return period. The estimation of catchment response times using empirical equations beyond their region of original development and without applying appropriate adjustment factors to compensate for the local differences in catchment characteristics, further exaggerates the situation. Hence, high runoff coefficients, underestimated time parameters and associated lower design rainfall depths, although of much higher intensities, would result in overestimated peak discharges, while low runoff coefficients and overestimated time parameters would result in underestimates.

5 CONCLUSIONS

The overall objective of this study was to present the further development and assessment of the DFET in 12 gauged catchments located in four climatologically different regions of South Africa. In applying the refined and improved DFET to the 12 gauged catchments, it proved to be an easy to use software tool for the rapid estimation and assessment of alternative design flood estimation methods for at-site applications in both gauged and ungauged catchments. In general, the empirical methods resulted in more acceptable design flood peaks than the deterministic methods when compared to the at-site probabilistic flood peaks. Furthermore, the use of more complex deterministic methods, e.g. SUH and LRH, as opposed to the more simplified deterministic methods, e.g. RM and SCS, does not necessarily guarantee and result in more acceptable peak discharge estimates. The latter also emphasises that no single design flood estimation method could be regarded as superior to any other method when the wide range of flood magnitude-frequency problems encountered in practice needs to be solved. Practitioners still have to apply their own experience, knowledge and skills to these particular problems until the current National Flood Study Programme (NFSP) [5] narrows the gap between flood research and practice in South Africa.

REFERENCES

[1] Pegram, G.G.S & Parak, M., A review of the regional maximum flood and rational formula using geomorphological information and observed floods. *Water SA,* **30**(3), pp. 377–392, 2004.

[2] Gericke, O.J. & Du Plessis, J.A., Development of a customised design flood estimation tool to estimate floods in gauged and ungauged catchments. *Water SA,* **39**(1), pp. 67–94, 2013. DOI: 10.4314/wsa.v39i1.9.

[3] Smithers, J.C., Review: Methods for design flood estimation in South Africa. *Water SA,* **38**(4), pp. 633–646, 2012.

[4] Van der Spuy, D. & Rademeyer, P.F., *Flood Frequency Estimation Methods as Applied in the Department of Water and Sanitation*, DWS: Pretoria, RSA, 2018.

[5] Smithers, J.C., Görgens, A.H.M., Gericke, O.J., Jonker, V. & Roberts, P., *The Initiation of a National Flood Studies Programme for South Africa*, SANCOLD: Pretoria, RSA, 2014.

[6] Van Dijk, M., *Utility Programs for Drainage (UPD)*, Sinotech CC: Pretoria, RSA, 2005. www.sinotechcc.co.za/Software/UPD/. Accessed on: 15 Mar. 2012.

[7] SANRAL, *Drainage Manual*, 6th ed., South African National Roads Agency Limited: Pretoria, RSA, 2013.

[8] Midgeley, D.C., Pitman, W.V. & Middleton, B.J., Surface Water Resources of South Africa, WRC Report No. 298/2/94. Water Research Commission: Pretoria, RSA, 1994.

[9] Gericke, O.J., Catchment response time and design rainfall: The key input parameters for design flood estimation in ungauged catchments. *Journal of the South African Institution of Civil Engineering,* **58**(1), pp. 2–15, 2018. DOI: 10.17159/2309-8775/2018/v60n4a6.

[10] DWAF, *GIS Data: Drainage Regions of South Africa,* Department of Water Affairs and Forestry: Pretoria, RSA, 1995.

[11] Smithers, J.C. & Schulze, R.E., Design Rainfall and Flood Estimation in South Africa, WRC Report No. 1060/01/03. Water Research Commission: Pretoria, RSA, 2003.

[12] Smithers, J.C. & Schulze, R.E., The estimation of design rainfall for South Africa using a regional scale invariant approach. *Water SA,* **30**(4), pp. 435–444, 2004.

[13] Gericke, O.J. & Du Plessis, J.A., Catchment parameter analysis in flood hydrology using GIS applications. *Journal of the South African Institution of Civil Engineering,* **54**(2), pp. 15–26, 2012.

[14] Midgeley, D.C. & Pitman, W.V., A Depth-Duration-Frequency Diagram for Point Rainfall in Southern Africa, HRU Report No. 2/78. University of the Witwatersrand, Hydrological Research Unit: Johannesburg, RSA. 1978.

DERIVATION OF THE INDEX GOVERNING THE SEEPAGE FLOW AND DYNAMIC SIMILARITY CONDITION OF LEVEE FAILURES DUE TO SEEPAGE FLOW

SHOJI FUKUOKA & KOSUKE TABATA
Research Development Initiative, Chuo University, Japan

ABSTRACT

It is necessary to establish an estimation method of levee weak points due to seepage flow during floods to make appropriate levee reinforcements and risk management measures in a river basin, preparing for large-scale floods. The internal flow with large phreatic surface gradient in the body of the levee tends to concentrate around the landside toe of the levee during a flood event, which may trigger the levee failure. This study shows that the idea of the levee failure is effective for primary selection of dangerous levee failure positions to the seepage flow. First, the only dynamic index governing the seepage flow in soils, the Seepage Flow number (SF_n), is derived by nondimensionalizing Richards' governing equation for a general seepage flow. Next, by applying SF_n to the seepage flow in the levee, the levee vulnerability index (t^*) is derived in the expression of a dynamic dimensionless quantity. Also, the vulnerability index of the levee on the permeable foundation is derived by using the unsteady quasi two-dimensional seepage flow analysis method. Finally, the vulnerability index (t^*) is calculated for the actual field levee and model levee experiment in which each failure mode such as leakage, sliding or collapse and breach occurred. As a result, it is confirmed that t^* shows approximately common values in the field levee and model levee taking the same failure mode. Therefore, it is demonstrated that t^* gives a dynamic similarity condition on the levee failure phenomena, due to the seepage flow and the judge of the degree of levee failures.

Keywords: seepage flow number, seepage line, levee failure, levee vulnerability index, dynamic similarity law, unsteady quasi two-dimensional equation, prototype and model.

1 INTRODUCTION

It is necessary to establish an estimation method of levee weak points due to seepage flow during a flood event to make appropriate levee reinforcements and risk management measures in a river basin. A general evaluation method of the levee safety was to calculate the danger of the slope sliding or collapse failure, for the steady seepage line assumed that a certain water level of the flood acts for a long time [1]. However, the internal flow with a large phreatic surface gradient in the body of the levee develops through the soil toward the landside toe during a flood event. The danger of levee failure will rise considerably if the seepage flow abnormally concentrates around the landside toe. Therefore, it will be possible to estimate when, where and what kind of levee failures will occur if the degree of concentration of the seepage flow at the landside toe can be appropriately evaluated.

In this study, the seepage phenomenon in the levee is expressed by the continuity equation and Darcy's law as the equation of motion. First, the index governing the seepage flow in the soil, the Seepage Flow number (SF_n), is derived by nondimensionalizing Richards' governing equation [2] for a general seepage flow. Next, by applying the SF_n to the seepage flow in the body of the levee, the levee vulnerability index (t^*) is derived in the expression of a dynamic dimensionless quantity. Moreover, it is confirmed that t^* gives a dynamic similar condition to the model and prototype levee. Also, each damaged level of the leakage, slope sliding or collapse and breach that t^* can take is discussed. This paper does not target the piping failure caused by the movement of water and the soil particle in a permeable foundation layer.

WIT Transactions on Ecology and the Environment, Vol 234, © 2019 WIT Press
www.witpress.com, ISSN 1743-3541 (on-line)
doi:10.2495/RBM190071

2 DIMENSIONLESS SIMILARITY INDEX GOVERNING GENERAL SEEPAGE FLOW AND LEVEE VULNERABILITY INDEX

2.1 Derivation of the similarity index governing the general seepage flow (SF_n)

First, we derive a general similarity index governing a seepage flow in soils, since the comparison of levee failures are made between field and model levees in following sections.

The research on the similarity law of the seepage flow has been conducted in the 1950–1960s. Bear [3] considered the similarity law of the steady seepage flow by non-dimensionalizing the flow continuity equation taking account into the seepage flow velocity expressed by Darcy's law. He proposed two conditions expressing the relationships among the dimensionless scales x', y', z' and dimensionless coefficients of permeability k_x', k_y', k_z'.

$$\frac{x'^2}{k_x'} = \frac{y'^2}{k_y'} = \frac{z'^2}{k_z'} \tag{1}$$

Corey et al. [4] suggested a similarity condition by nondimensionalization of Richards' equation [2]. However, it was a study focusing on the vertical seepage flow under the gravitational drainage in the saturated ground, and the pressure head and seepage length were given as one common length. Therefore, the suggested dimensionless value could not be a similarity law of the general seepage flow. Although other several similarity relations [5], [6] in which were expressed by empirically combining the length, permeability and porosity were suggested, the verifications were not enough performed. There is no similarity condition for the seepage flow with generality as far as the authors know.

The governing equation of the seepage flow in the soil is expressed by Richards' equation (eqn (2)),

$$\frac{1}{\rho}\frac{\partial \rho \lambda S_r}{\partial t} - \frac{\partial}{\partial x}\left(k\frac{\partial h}{\partial x}\right) - \frac{\partial}{\partial y}\left(k\frac{\partial h}{\partial y}\right) - \frac{\partial}{\partial z}\left(k\frac{\partial h}{\partial z}\right) = 0, \tag{2}$$

where ρ is density of water, λ is porosity of soil, S_r is degree of saturation, k is coefficient of permeability, h is total head ($=\Psi+z$), Ψ is pressure head. It is necessary to hold the same equation between the prototype and the model by nondimensionalizing the governing equation for deriving the dynamic similarity law. The dimensionless value of each variable of eqn (2) is given as follows by using each representative value. Here, superscript ' and subscript 0 mean dimensionless and representative quantity, respectively.

$$\rho' = \frac{\rho}{\rho_0}, S_r' = \frac{S_r}{S_{r0}}, \lambda' = \frac{\lambda}{\lambda_0}, h' = \frac{h}{H_0}, t' = \frac{t}{T_0}, k' = \frac{k}{k_0}, x' = \frac{x}{L_0}, y' = \frac{y}{L_0}, z' = \frac{z}{L_0} \tag{3}$$

When eqn (2) is arranged taking account into eqn (3), the following dimensionless equation is finally obtained.

$$\frac{1}{\rho'}\frac{\partial \rho' \lambda' S_r'}{\partial t'} + \boxed{\frac{1}{S_{r0}}\frac{k_0 H_0 T_0}{\lambda_0 L_0^2}}\left\{-\frac{\partial}{\partial x'}\left(k'\frac{\partial h'}{\partial x'}\right) - \frac{\partial}{\partial y'}\left(k'\frac{\partial h'}{\partial y'}\right) - \frac{\partial}{\partial z'}\left(k'\frac{\partial h'}{\partial z'}\right)\right\} = 0 \tag{4}$$

In this dimensionless eqn (4), the dimensionless value enclosed with a red frame ahead of parentheses that settle the second, third and fourth terms in the left side appears. This dimensionless value is defined as SF_n, which shows the dynamic similarity condition of the seepage flow. If the seepage flow in the field (prototype) and the model have the same value of SF_n, the seepage flow of both will be dynamically similar.

2.2 Derivation of levee vulnerability index t^* from SF_n

The presented SF_n is expressed in the ratio of the time of $S_{r0}\lambda_0 L_0^2/k_0 H_0$ to T_0. The former indicates the time required to move the representative horizontal distance L_0 by the true seepage flow velocity $k_0 H_0/S_{r0}\lambda_0 L_0$. The latter indicates the representative time of the seepage flow.

The SF_n can be rewritten in the index that relates the seepage flow in the levee to the levee failure if each representative physical value that consists SF_n is chosen properly so as to describe the seepage phenomena in the levee.

By substituting H, b, t', k, λ and S_r to H_0, L_0, T_0, k_0, λ_0 and S_{r0}, SF_n is finally shown by eqn (5). Here, H is water level, b is horizontal length from water front to levee toe of landside, t' is duration time from when the water level exceeds the ground level of the landside toe. k, λ and S_r are averaged permeability, porosity and degree of saturation of the levee (see Fig. 1). Then, SF_n is defined as levee vulnerability index t^*.

$$t^* = \frac{1}{S_r}\frac{kHt'}{\lambda b^2} \tag{5}$$

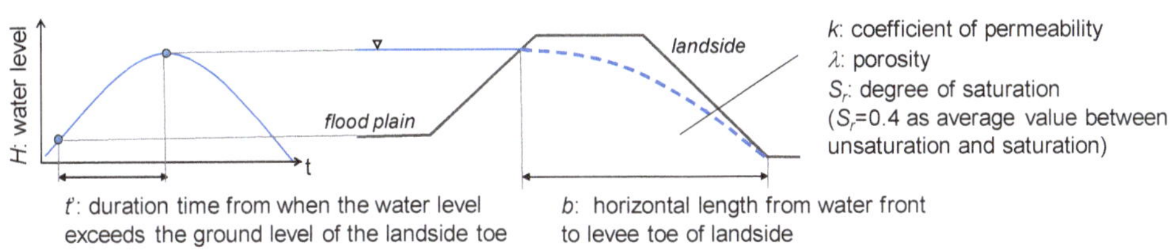

Figure 1: Definition of variables of levee vulnerability index t^*.

2.3 Derivation of t^* from quasi two-dimensional unsteady seepage flow equation

Fukuoka and Tabata [7] assumed the seepage line in the levee on the permeable foundation as the similar function form (eqn (6)).

$$h(x,t) = H\left\{1-\left(\frac{x}{\xi(t)}\right)^m\right\}, \tag{6}$$

where ξ is a front position of seepage line, m is a parameter which defines the shape of the seepage line.

In addition, a differential equation of quasi two-dimensional unsteady seepage flow (see Fig. 2) that can describe the movement of the seepage line was derived based on Darcy's law and continuity condition of water in the levee and foundation. The derived equation was solved for the front position ξ of the seepage line and let T be the time required for the seepage line to reach the horizontal distance b. Then, the general levee vulnerability index (eqn (7)) defined by $t^*_m = t'/T$ was obtained.

$$t_m^* = \frac{2(m+1)}{2m-1}\frac{mk_1 H - m(2m-1)k_2 D}{mH+(m+1)D}\frac{Ht'}{\lambda b^2}, \tag{7}$$

where, k_1 and k_2 are the coefficients of permeability of the levee and foundation, D is the thickness of the permeable layer and defined as the distance from the impermeable layer or ground water level to the ground level under the levee. The parameter m in eqn (6) can be decided by Table 1, based on the values of k_2/k_1 and D/H of the levee and foundation [7].

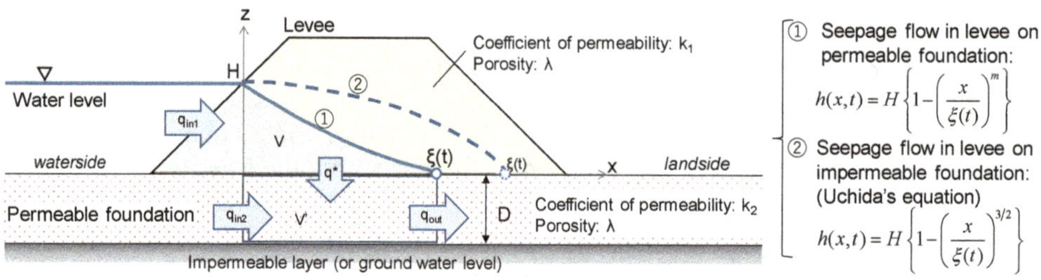

Figure 2: Quasi two-dimensional unsteady seepage flow in the levee [7].

Table 1: Relationship of k_2/k_1, D/H and parameter m [7].

D/H \ k_2/k_1	0	0.1	0.5	1	2	5	10	20
0	1.500	1.500	1.500	1.500	1.500	1.500	1.500	1.500
0.1	1.500	1.500	1.500	1.500	1.426	1.060	0.858	0.711
0.5	1.500	1.500	1.472	1.075	0.823	0.660	0.588	0.545
1	1.500	1.500	1.118	0.860	0.700	0.591	0.546	0.524
2	1.500	1.384	0.866	0.710	0.614	0.547	0.524	0.510
5	1.500	1.014	0.671	0.592	0.549	0.519	0.509	0.504
10	1.500	0.821	0.591	0.548	0.524	0.509	0.504	0.501
20	1.500	0.691	0.548	0.524	0.510	0.504	0.501	0.501

In the case of the levee on the impermeable foundation, D and k_2 in eqn (7) are zero. Also, the parameter m is chosen 3/2 in order to coincide with the one-dimensional unsteady seepage line equation suggested by Uchida [8]. Then, $2(m+1)/(2m-1)$ in eqn (7) becomes the value of 5/2. As a result, the levee vulnerability index t^* of the levee on the impermeable foundation is given by eqn (8).

$$t^* = \frac{5}{2}\frac{Hk_1t'}{\lambda b^2} \tag{8}$$

This equation means that the true seepage flow velocity in the levee is described by $5/2Hk_1/\lambda b$ when the water level in the river H acts on the unsaturated levee with width b, coefficient of permeability k_1 and porosity λ. Also, it can be interpreted that the influence of degree of saturation S_r in eqn (5), which changes as the seepage line develops, is included in the coefficient 5/2.

Uchida [8] expressed the averaged velocity of the seepage flow in the levee as $8/3Hk_1/\lambda b$ by solving the temporary change of the seepage line based on the graphical analysis method and confirmed it by seepage flow experiments. Therefore, as the seepage line develops and the effects of saturation and unsaturation in the soil are integrated, finally the moving speed of the seepage line is described by $8/3Hk_1/\lambda b$ on average. Since the degree of saturation S_r of soil is about 0.3 in the completely gravity drained state and 1 in the saturated state, $1/S_r$ of t^*

(eqn (5)) derived from SF_n takes about 1 to 3. Therefore, it is can be regarded that eqns (5) and (8) almost coincide. In general, it is known that the permeability coefficient varies according to the degree of saturation, but this paper does not refer to how to incorporate the change in saturation at unsteady seepage phenomena and the corresponding permeability coefficient.

In the case of the levee on the permeable foundation, the time T required for the seepage line to reach the landside toe will be larger compared with the case of the levee on the impermeable one, because a part of water in the levee flows into the permeable foundation and it makes the seepage line lower. For these reasons, the levee vulnerability index t^*_{m} on the permeable foundation is calculated smaller than t^* on the impermeable one. Therefore, the levee vulnerability index t^* that assumes the impermeable foundation should be applied as the index of a dangerous failure.

The levee vulnerability index t^* on the impermeable foundation shown by eqn (8) is described in the same formula form with index (eqn (5)) obtained by applying SF_n to the seepage flown in the levee. This means that the levee vulnerability index t^* is the only dynamic similarity condition of the seepage flow in the levee.

3 APPLICATION OF LEVEE VULNERABILITY INDEX

3.1 Application of the levee vulnerability index t^* to field levees

The levee vulnerability index t^* of the field levees in the Kinu River 2015 flood, the Koyoshi River 2013 flood, the Kakehashi River 2012 flood, the Yabe River 2012 flood and the Nagara River 1976 flood is calculated. In these rivers, the levee damage occurred considerably due to seepage flows during flood events (see Fig. 3(a) and (b)).

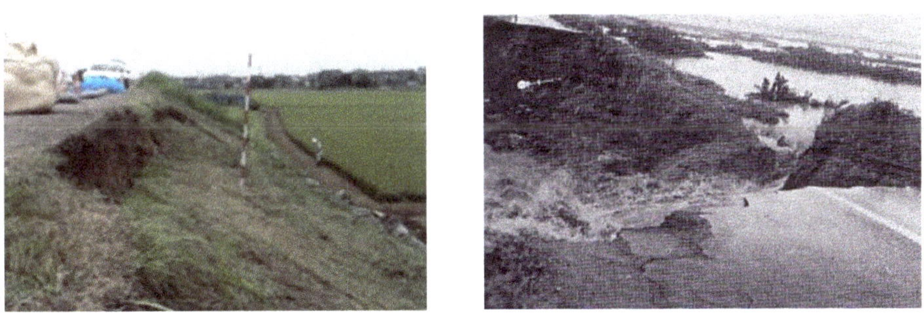

(a) (b)

Figure 3: The levee failure due to the seepage flow observed in the field levee. (a) Sliding at 8.4 k right bank in the Kakehashi river during the 2012 flood; (b) Levee breach at 33.8 k right bank in the Nagara river during the 1976 flood.

Fig. 4 shows the value of the levee vulnerability index of the respective field levee calculated by using eqns (7) and (8). The coefficient of permeability is estimated based on Hazen [9] or Creager et al.'s [10] method, by using D_{10} and D_{20} observed by the boring survey data of the levees and larger one of Hazen and Creager's method is given to the calculation. The calculated t^* are plotted with larger variation compared with the model levee experiments stated in 3.2 since the influences of rainfall and slope protection are not taken into consideration. However, the levee vulnerability index t^* of the field levee in which the levee breach (red circles in Fig. 4) and sliding or collapse (yellow circles in Fig. 4) occurred take approximately 0.1–1.0 and 0.01–0.1 respectively. In which the leakage from the levee (without circles in Fig. 4) occurred, the variation of the data is large, but most of the t^* are distributed in the range larger than 0.001.

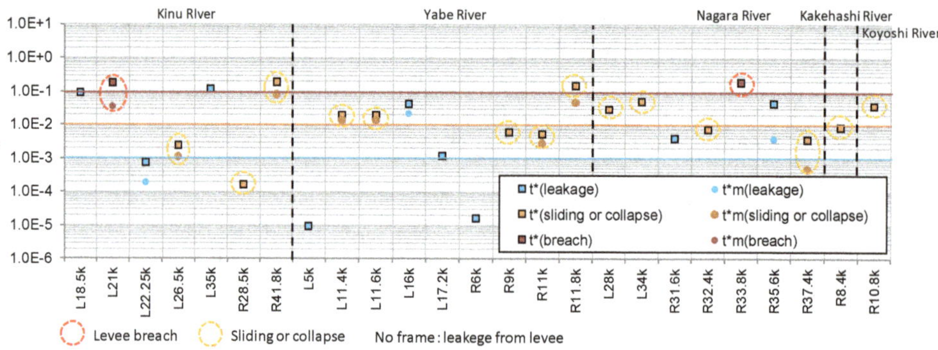

Figure 4: The values of the levee vulnerability index t^* of the field levees.

3.2 Application of the levee vulnerability index t^* to model levee experiments

Yoshikawa and Kodaka [11] conducted a series of model levee experiments for clarifying the relationship between the soil structure and the levee failure mechanism due to the seepage flow. Three kinds of models such as two-dimensional half cross section, two-dimensional full cross section and three-dimensional full cross section were created, and a total of 22 cases of experiments whose thickness and permeability of the foundation layer were different were carried out. Fig. 5 shows the specification of the two-dimensional full section model and example of the levee failure due to the seepage flow observed in the model levee. The model levee is 0.2–0.3 m in height and 0.7–0.9 m in width. The foundation is created by the different permeable materials in the upper and lower layer. The outside water level was rapidly increased from the initial water level condition while the lower layer was saturated, then the water level was kept constant. The deformation process of the body of levees and its occurrence time were observed.

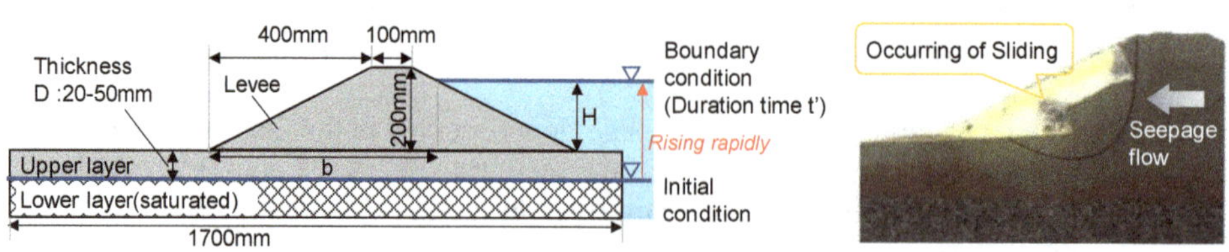

Figure 5: Specification of the two-dimensional full section model levee and example of the levee failure due to the seepage flow observed in the model levee.

By using these experimental data, the relationship between t^* and the failure modes is investigated. Fig. 6 shows the values of the levee vulnerability index t^* of the model. The levee vulnerability index t^* of the experiments in which the levee breached are almost all plotted in the range of 0.1–1.0. In which the sliding or collapse occurred, t^* are concentrated at the range of 0.01–0.1 and plotted 16 data at this range. In addition, most of t^* calculated for cases which the leakage from the levee are observed are approximately plotted in the range of 0.001 to 0.01. These results are the same as ones of the field levees in 3.1.

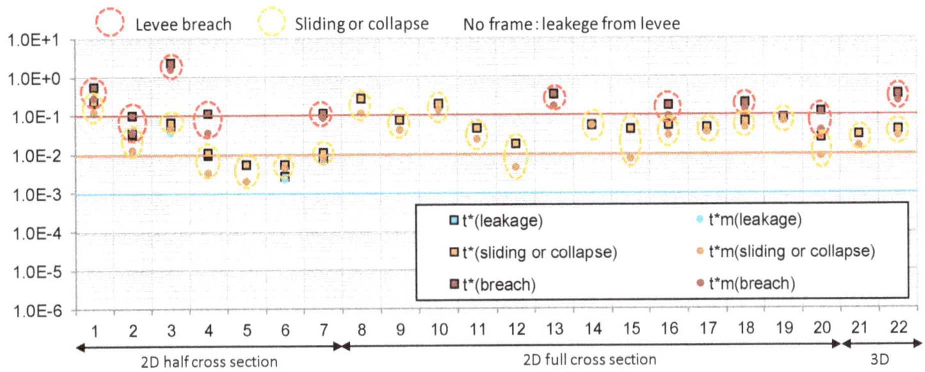

Figure 6: The values of the levee vulnerability index t^* of the model levees.

Consequently, t^* can approximately elucidate the degree of the levee damage and the danger due to the seepage flow. It is clear that t^* is the dynamic similarity condition established commonly for model levees and field levees since the range of t^* taken by each failure mode in the field levee and model levee experiment is almost in agreement.

4 CONCLUSION

Based on the idea that the internal flow with large phreatic surface gradient in the body of levees concentrates around the landside toe of the levee during a flood event, which may trigger the levee failure, the estimation method for the degree of the levee failure due to the seepage flow is presented. First, the only dynamic index governing the seepage flow in the soil, SF_n, is derived by nondimensionalizing Richards' governing equation for a general seepage flow. Next, by applying SF_n to the seepage flow in the levee, levee vulnerability index t^* is derived in the expression of a dynamic dimensionless quantity. Also, the vulnerability index of the levee on the permeable foundation is derived by using unsteady quasi two-dimensional seepage flow analysis method. Finally, the vulnerability index t^* is calculated for the actual field levee and model levee experiment in which each failure mode such as leakage, sliding or collapse and breach occurred. As a result, it is confirmed that t^* shows approximately common values in the field levee and model levee experiment taking the same failure mode. Therefore, it is demonstrated that t^* gives a dynamic similarity condition on the levee failure phenomena due to seepage flow. Also, it is shown that the degree of levee failure can be judged by t^* value.

ACKNOWLEDGEMENT

In this study, we used valuable data of the seepage flow in the model levee experiment conducted by the research group of Professor Takeshi Kodaka of Meijo University, Japan. We would like to express our appreciation.

REFERENCES

[1] CIRIA, *The International Levee Handbook*, C731, London, 2013.

[2] Richards, L.A., Capillary conduction of liquids through porous mediums. *Journal of Applied Physics*, **1**, pp. 318–333, 1931.

[3] Bear, J., *Dynamics of Fluids in Porous Media*, American Elsevier, pp. 671–677, 1972.

[4] Corey, G.L., Corey, A.T. & Brooks, R.H., *Similitude for Non-Steady Drainage of Partially Saturated Soils,* Hydrology Papers 9, Colorado State University, 1965.

[5] Kawaharada, R., *On the Hydraulic Similarity of the Unsteady Seepage Flow*, Faculty of Agriculture, Kagoshima University, vol. 5, pp. 115–117, 1956 (Japanese).

[6] Nakamura, M., Fundamental theory of motion of percolation water (IV) A theory of model experiment. *Journal of the Agricultural Engineering Society*, **25–26**, 1957 (Japanese).

[7] Fukuoka, S. & Tabata, K., Development of seepage line estimation method considering permeable foundation ground under the levee and reinforcing against the levee failure. *Advances in River Engineering*, **22**, pp. 261–266, 2016 (Japanese).

[8] Uchida, S., On the non-steady percolation with a free boundary. *Japan Society of Civil Engineers*, pp. 58–62, 1952 (Japanese).

[9] Hazen, A., Some physical properties of sands and gravels with special reference to their use in filtration, Massachusetts State Board of Health, 24th Annual Report, pp. 539–556, 1892.

[10] Creager, W.P., Justin, J.D. & Hinds, J., *Engineering for Dams,* vol. III, John Wiley & Sons, 1945.

[11] Yoshikawa, T. & Kodaka, T.A., Consideration of the failure mechanism of river levees due to water seepage based on model tests and soil-water-air coupled finite deformation analysis. *Proceedings of the 19th International Conference on Soil Mechanics and Geotechnical Engineering*, pp. 1319–1322, 2017.

ESTIMATION OF LEVEE FAILURE POINTS BASED ON LEVEE VULNERABILITY INDEX AND FLOOD RISK ANALYSIS BY INTEGRATING SIMULATION OF FLOOD FLOW AND INUNDATION

KOSUKE TABATA & SHOJI FUKUOKA
Research Development Initiative, Chuo University, Japan

ABSTRACT

To examine the flood risk management measures, it is important to estimate the levee breach points, inundation discharge hydrograph and motion of the inundation water in the basin as precisely as possible. The purpose of this paper is to conduct the above series of studies on the Kinu River (Japan) suffered from inundation in the range of 40 km^2 due to overflowing and levee breach during the 2015 large flood. First, a simulation model integrating flood flow and inundation is developed. The inundation discharge hydrographs due to the overflowing and levee breach are estimated by the flood flow analysis based on observed temporal changes in water surface profiles in the Kinu River. The inundation is calculated by a two-dimensional model. The mesh of the inundation analysis model is generated by regular grid with 5 m size in order to express the microtopography obtained by DSM (Digital Surface Model). Manning's roughness coefficients are given according to the situation of the land use and inundation depth. It is confirmed that the developed model is useful by the inundation arrival time read from camera images and the spread of the actual inundation water in the basin. Next, the suitability of the levee vulnerability index t^* which has been derived by the authors is examined for the levees of the Kinu River. Finally, the inundation flow analysis is conducted under the actual and hypothetical levee breach conditions based on the value of t^*. It is concluded that the difference in the location and time of the levee breach influences on the motion of the inundation water and inundation areas in the Kinu River basin.

Keywords: Kinu River 2015 flood, inundation flow, flood flow, levee vulnerability index, flood risk.

1 INTRODUCTION

It is important to estimate the levee breach points, inundation discharge hydrograph and motion of the inundation water in the basin as precisely as possible to examine the flood risk management measures.

Tabata et al. [1] developed a flood flow analysis model based on the temporary changes of observed water surface profiles during the 2015 flood in the Kinu River, Japan, which suffered from damage in the huge inundation area due to levee breach. The inundation discharge hydrographs were estimated accurately by the developed flood flow model. Also, in the Joso City which faces the left bank of the Kinu River and inundated in the 2015 flood, the inundation time and depth were obtained from camera images installed at some locations in the Kinu River basin. These data will make it possible to verify the inundation analysis model.

In general flood risk assessment [2], levee breach points are usually set up at levee failure points where the water level exceeds the design water level. However, serious levee damage such as a levee breach and sliding or collapse has occurred at points where the body of the levee is vulnerable to seepage flow even in the section where the water level is below the design water level. In recent years, Fukuoka and Tabata [3] have proposed the levee vulnerable index t^* which is represented by the combination of water level, duration time of flood, levee width, porosity and permeability of the levee. It has been confirmed

www.witpress.com, ISSN 1743-3541 (on-line)
doi:10.2495/RBM190081

that the levee vulnerability index can represent the risk of levee failure due to seepage flow. By using this index, the levee breach points will be estimated rationally.

The purpose of this paper is to estimate the flood risk by integrating simulation of flood flow and inundation in the Kinu River basin. First, the inundation flow analysis model is developed by using the inundation discharge hydrographs due to the overflowing and levee breach. Then, the inundation model is verified by the filed inundation data. Next, the levee failure risk due to seepage flow in the Kinu River is evaluated by the use of the levee vulnerability index t^*. Finally, the inundation flow analysis is conducted under the levee breach condition based on the value of t^*. It is concluded that the difference in the location and time of the levee breach considerably influences on the motion of the inundation water and inundation area in the Kinu River basin.

2 DEVELOPING THE INTEGRATING SIMULATION MODEL OF FLOOD FLOW AND INUNDATION

2.1 Integrated simulation model of flood flow and inundation in the Kinu River

Fukuoka and Watanabe [4] have proposed the numerical method for flood flows and bed variations using time series of observed flood water surface profiles on the basis of the idea that effects such as plan forms, cross-sections and outflowing water from the river appear in time series of observed water surface profiles of flood flows. Based on this idea, the integrated simulation model of flood flow and inundation was developed by using time series of the observed water surface profiles during the 2015 flood in the Kinu River [1]. Since overflowing and collapse of the levee occurred around 25.35 k, the inundation flow was calculated so that the collapse occurred when the water level exceeded the levee height. Also, the inundation flow from the levee breach point at 21.0 k was calculated so that the calculated water surface profiles agreed with the observed ones around the levee breached point [5] considering observed breach time and temporary change of breach width (Fig. 1). This model could not only elucidate the flood propagation in the Kinu River but also estimate inundation discharge hydrographs due to the overflowing and levee breach.

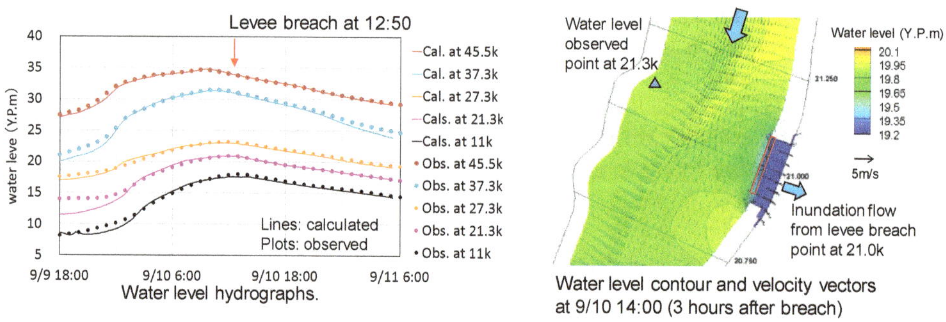

Figure 1: Calculated water level and inundation flows in the levee breach point at 21.0 k.

Fig. 2 shows the inundation discharge hydrographs which were estimated by this model. The first wave was caused by overflowing and collapse of levee around 25.35 k, which started at 06:00 on 10 September 2015, and the second one was caused by the levee breach at the 21 k left bank at 12:50 on 10 September 2015. The calculated inundation discharge exceeded 500 m^3/s at the maximum and the volume was estimated to be 32,000,000 m^3. The measured total inundation volume was about 38,000,000 m^3, which agreed well with the computed result. The inundation flow analysis is carried out by using these inundation discharge hydrographs as the boundary conditions.

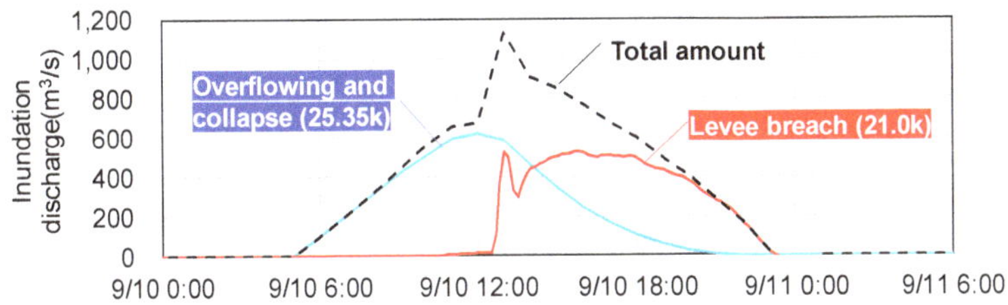

Figure 2: Estimated inundation discharge during the 2015 flood event in the Kinu River.

2.2 Inundation flow analysis in Joso City

A two-dimensional inundation analysis model for Joso City which faces the left bank levee in the Kinu River is developed by using the inundation discharge during the 2015 flood estimated by the analysis model in Section 2.1.

Fig. 3 shows the developed model for inundation analysis model in Joso City. The mesh is generated by regular grid with 5 m size in order to express the microtopography obtained by DSM (Digital Surface Model). Also, the mean ground level of each mesh is given by DSM. The land use classifications were set based on GIS data of farmlands, road networks and water area published by the Geophysical Survey Institute of Japan, and they were given to each mesh. Manning's roughness coefficients are given according to the land use and inundation depth. The roughness coefficients of farmland and water areas are given as 0.070 and 0.020 respectively. The roughness of the road is given small as 0.025 since it is generally known that the inundation water flows with high velocity on the road [6]. A large resistance acts on the inundation front flowing into the dry area, and the resistance is reduced at the place where was wet once. Therefore, a depth threshold h_0 (1.0 m in this study) related to dry or wet state is introduced. If the water depth is less than h_0 or has never been flooded, the roughness coefficient is given as 0.100, otherwise the one for each land use is given.

Fig. 4 shows a comparison between the observation and calculation of the temporal changes of the inundation depth at observed points St.1 and 2 in Fig. 1. The observed data were read from camera images at St.1 and 2. It can be confirmed that the inundation starts immediately after the levee breach at St.1. Also, the calculated data can roughly explain the observed one at St.2 which is far from the site of overflowing and levee breach. Therefore, it is shown that the developed model can express the spread of the inundation water from the Kinu River and the distribution of inundation water depth in Joso City.

Fig. 4 shows the inundation depth contours at 15:00 on 10 September and 0:00 on 11 September. First, part of the inundation water around 25.35 k flows into the main drainage channel and reaches the downstream part via the channel. At this time, the inundation water due to the levee breach at 21.0 k left bank was combined with the inundation water due to overflowing and collapse around 25.35 k. Also, the main body of the inundation water reached near the Joso City hall at 0:00 on 11 September (Fig. 4(b)), and a large-scale inundation occurred around the Joso City hall.

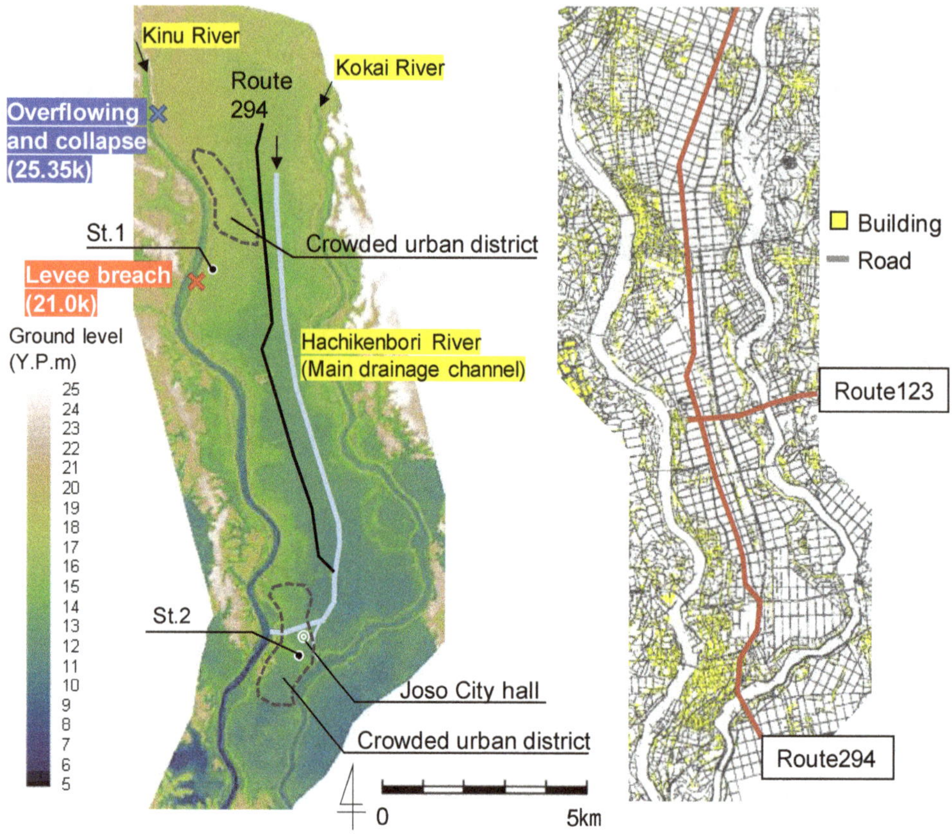

Figure 3: Developed inundation analysis model in Joso City.

3 EVALUATION OF LEVEE FAILURE RISK AND LEVEE BREACH POINTS BASED ON LEVEE VULNERABILITY INDEX

3.1 Levee vulnerability index t^* by the seepage flow

Fukuoka and Tabata [3] have proposed the dimensionless index for the levee failure due to seepage flow during a flood event; levee vulnerability index t^* (eqn (1)).

$$t^* = \frac{5}{2}\frac{kHt'}{\lambda b^2}. \tag{1}$$

Here, H is water level, b is horizontal length from the water front of waterside levee to the toe of the landside levee, t' is duration time from when the flood water level exceeds the ground level of the landside toe. k, λ and S_r are averaged permeability, porosity and degree of saturation of the levee, respectively (see Fig. 5(a)).

The levee vulnerability index t^* has been applied to many field levees where the damage had occurred, and it has become clear that when the value of t^* exceeds 0.1, the risk of levee breach due to the seepage flow tends to increase [3].

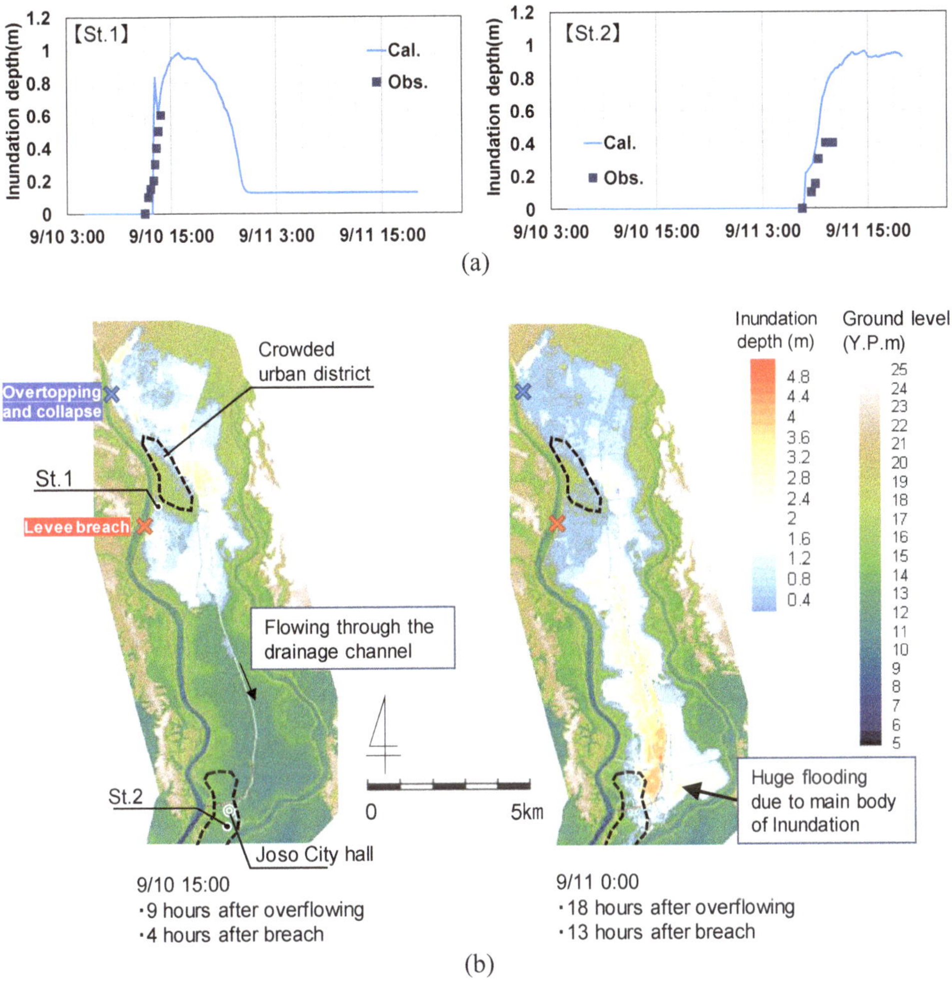

Figure 4: Inundation depth distribution in the Kinu River basin. (a) Comparison between calculation and observation at St.1 and St.2; and (b) Calculated inundation depth contour.

When the water level recedes, the top of the seepage line in the levee gradually moves to the landside toe. Therefore, the seepage flow from the top of the seepage line to the landside toe of the levee keeps on going and maintaining a high risk of levee failure. This is mainly caused by the reduction of the horizontal length b in eqn (1) due to the shift of the seepage peak height. Uemura and Fukuoka [7] have proposed the analysis method that can accurately represent the above-mentioned unsteady seepage phenomenon during the flood water receding stage. The outline of this method is shown below.

Consider the continuity condition that the volume change of water in the levee $\lambda\delta V$ coincides with the displacement $q\delta t$ from the body of the levee to the waterside due to the water level receding (see Fig. 5(b)).

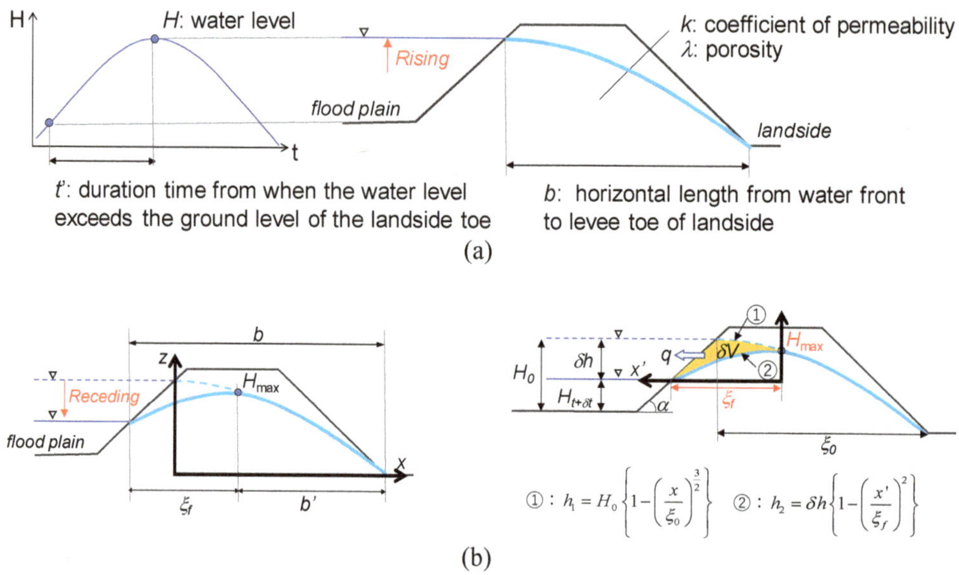

Figure 5: Definition of variables of levee vulnerability index t^*. (a) During water level rising (definition of variables in eqn (1)); and (b) During water level receding (definition of variables in eqns (2)–(5)) [7].

$$\lambda \delta V = q \delta t \,. \tag{2}$$

Here, δt is seepage time, δV is volume change of water in the levee per δt, q is outgoing discharge from waterside slope of the levee as the water level of rivers recedes. The left side of eqn (2) is indicated by the orange hatched area in Fig. 5(b) and expressed as eqn (3).

$$\begin{aligned}\lambda \delta V = \frac{\lambda \delta h^2}{2\tan\alpha} - \frac{2}{3}\lambda \xi_f \left\{ \delta h - \left(\delta h + H_{t+\delta t}\right)\left(\frac{-\xi_f + \delta h/\tan\alpha}{-\xi_0}\right)^{3/2} \right\} \\ -\lambda\left(-\xi_f + \frac{\delta h}{\tan\alpha}\right)\left\{\frac{2}{5}\left(\delta h + H_{t+\delta t}\right)\left(\frac{-\xi_f + \delta h/\tan\alpha}{-\xi_0}\right)^{3/2} - \delta h\right\}.\end{aligned} \tag{3}$$

The right side of eqn (2) is expressed as eqn (4) by using Depuit–Forchheimer's equation [8]

$$q\delta t = \frac{k}{2\xi_f}\left[\left[H_0\left\{1 - \left(\frac{-\xi_f + \delta h/\tan\alpha}{-\xi_0}\right)^{3/2}\right\}\right]^2 - H_{t+\delta t}^2\right] * \delta t \,. \tag{4}$$

The horizontal position of seepage line's top ξ_f (see Fig. 5(b)) can be numerically solved since eqn (3) is equal to eqn (4). Then, the peak height of the seepage line's top H_{max} will be found by assuming that the position of the peak of the water level shifts along the seepage line. Therefore, the levee vulnerability index during the water level receding stage is expressed as eqn (5) by substituting H_{max} and b' ($=b-\xi_f$) for H and b in eqn (1)

$$t^* = \frac{5}{2}\frac{kH_{max}t'}{\lambda b'^2}. \tag{5}$$

3.2 Calculated levee vulnerability index t^* and estimating of levee breach points

The levee vulnerability index t^* was calculated longitudinally for the Kinu River's left bank which faces Joso City. Fig. 6 shows the longitudinal distribution of the levee vulnerability index t^* at peak water level (black plots) and the maximum value of t^* during the water level receding stage (red plots). t^* around the actual breach point (21.0 k) are plotted in the range of 0.1 to 1.0 where the levee breach risk is very high. Besides these, t^* around 11.75 k exceed 0.1. Therefore, if the levee breach did not occur at 21.0 k, it might have occurred around 11.75k.

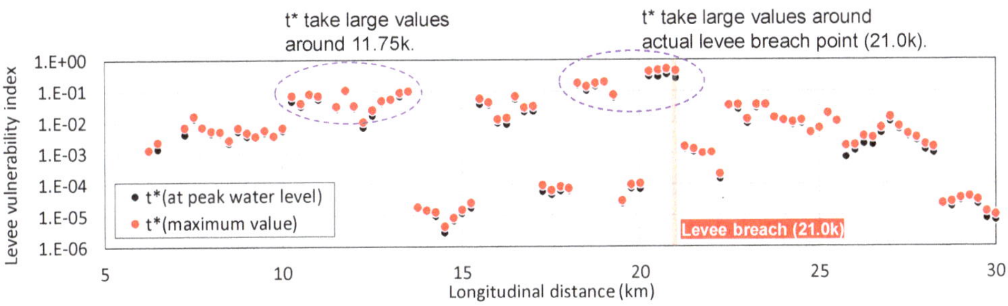

Figure 6: Longitudinal distribution of t^* along the left bank of the Kinu River.

Fig. 7 shows the temporary change of the water level and levee vulnerability index at 21 k and 11.75 k. It is confirmed that t^* exceeds 0.1 around the actual levee breach time (12:50 on 10 September) at 21 k left bank. Assuming that the levee breach occurs when t^* exceeds 0.1, the levee breach time at 11.75 k can be estimated as 17:00 on 10th September which is two hours after the peak water level.

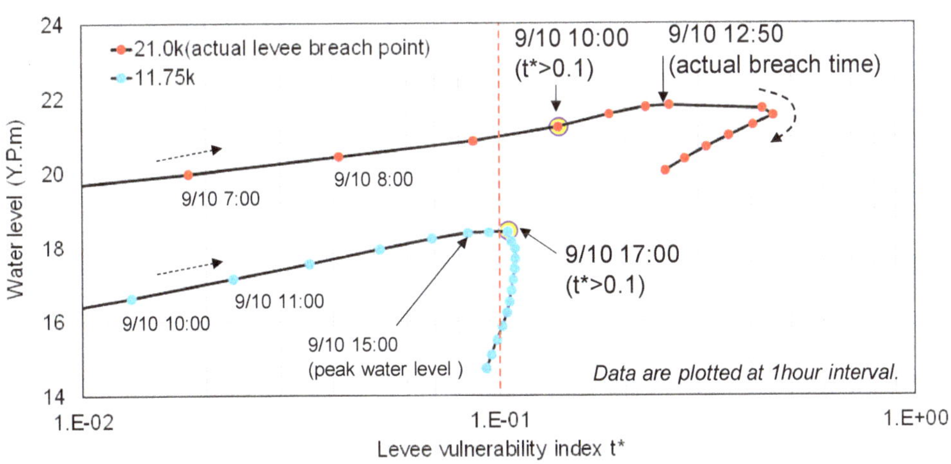

Figure 7: Temporary changes of water level and t^* at 21.0 and 11.75 k.

4 INUNDATION ANALYSIS UNDER A LEVEE BREACH SCENARIO BASED ON LEVEE VULNERABILITY INDEX

We need to conduct the inundation analysis under levee breach scenarios different from the actual flood inundation stated in Section 2.2 to investigate the inundation risk in the basin. In this section, the inundation discharge hydrograph is estimated, and the inundation analysis is performed under a hypothetical levee breach condition based on the result of the levee vulnerability index.

The hypothetical levee breach point was set at 11.75 k left bank in which t^* value took the second largest value after the actual levee bleach point at 21.0 k (see Fig. 6). Also, the levee breach time was set at 17:00 on 10 September when t^* exceeds 0.1 (see Fig. 7). This analysis was not taking account into the overflowing and collapse around 25.35 k. Fig. 8 shows the inundation discharge hydrograph due to the levee breach at 11.75 k left bank estimated by the simulation integrating the flood flow and inundation. The inundation discharge becomes about 500 m^3/s as well as the actual breach point at 21.0 k. However, since the levee breaches during the flood receding stage, the inundation volume is 7,500,000 m^3, which is smaller than the actual one.

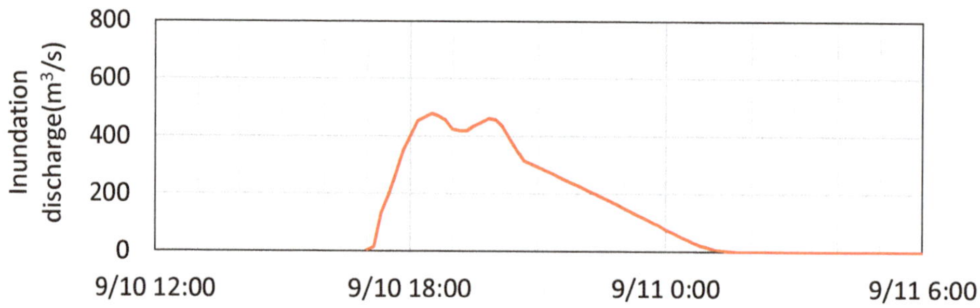

Figure 8: Estimated inundation discharge hydrograph due to the levee breach at 11.75 k.

The inundation flow analysis for the hypothetical levee breach condition at 11.75 k left bank based on the value of t^* was conducted. Fig. 9 shows the time change of the inundation depth contour. Although the inundation area is smaller than the actual one (Fig. 4), the inundation water spreads rapidly because the levee breach point is close to the crowded urban district of Joso City. It is shown that the maximum inundation depth in the crowded urban district (black dashed circle in Fig. 9) rises over 2.5 m, 4 hours after the levee breach at 11.75 k. Then, much of the inundation water flows northward on the west side of the main drainage channel. 12 hours after the levee breach, inundation water also spreads widely to the east of the main drainage channel. It is clear the crowded urban district of Joso City would suffer from the comparable large flood damage even if either 21k or 11.75k levee breached.

5 CONCLUSION

The integral simulation model for flood flow and inundation in the Kinu River basin suffered from inundation due to the levee breach at 21.0 k and overflowing and collapse around 25.35 k during the 2015 large flood was developed. This model can approximately elucidate the inundation arrival time read from camera images and the actual inundation water spread in the basin.

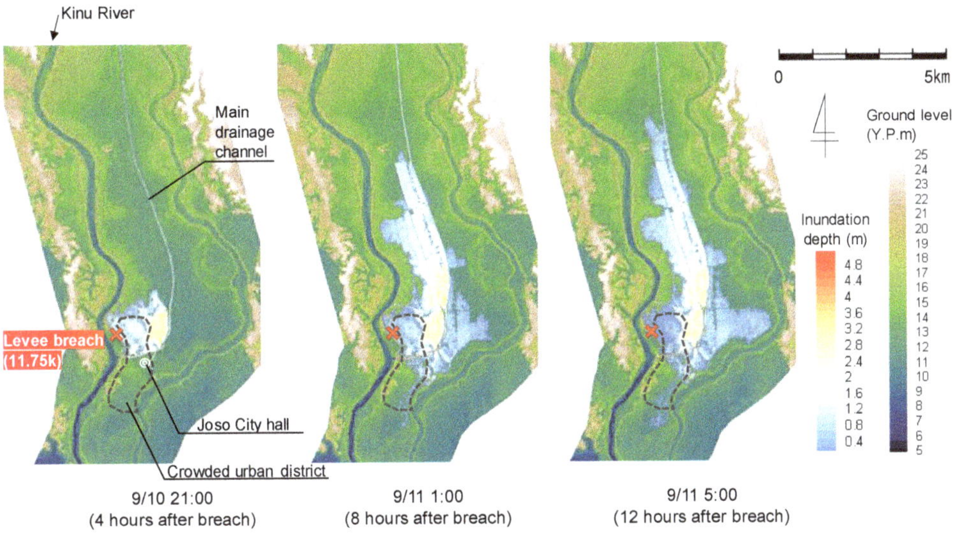

Figure 9: Inundation depth contour which is calculated taking account into the levee breach at 11.75 k left bank in the Kinu River.

Also, the levee vulnerability index t^* was calculated along the levee in the Kinu River and took high values at not only the actual levee breach point in 21.0 k but also 11.75 k.

The hypothetical inundation flow analysis by assuming that the levee breach occurs at 11.75 k was conducted. It is clear that the south crowded urban district of Joso City would suffer from the comparable huge flood damage even if either 21 k or 11.75 k levee breached. Therefore, the levee reinforcement against the seepage flow is needed at the places where the levee vulnerability index t^* take high values. It is expected to make an advanced inundation hazard map by performing the inundation flow analysis under the several levee breach scenarios based on the levee vulnerability index t^*.

REFERENCES

[1] Tabata, K., Fukuoka, S. & Deguchi, K., Analysis of discharge capacity and flood storage rate during 2015 large flood in the Kinu River. *12th International Conference on Hydro-Science and Engineering for Environmental Resilience, ICHE 2016*, 15-0036, 2016.

[2] Bilal, M., Jerry, F. & William, L., Risk analysis of a protected hurricane-prone region. I: model development. *Natural Hazards Review*, **10**(2), pp. 38–53, 2009.

[3] Fukuoka, S. & Tabata, K., Index governing the seepage flow and dynamic similarity condition of levee failure due to seepage flow: Seepage flow number and levee vulnerability index. *Annual Journal of Hydraulic Engineering, JSCE*, **74**(5), pp. I_1435–1440, 2018. (In Japanese.)

[4] Fukuoka, S. & Watanabe, A., Estimating Channel Storage and Discharge Hydrographs of Flooding. *Proceedings of the 5th International Conference on Hydro-Science and Engineering*, 2002.

[5] Fukuoka, S., Abe, T. & Tsukamoto, Y., Calculation method of inundation discharge hydrograph from the temporal changes in water surface profile due to levee breach. *Journal of JSCE*, **65**(3), pp. 166–178, 2009. (In Japanese.)

[6] Fukuoka, S., Kawashima, M., Matsunaga, N. & Maeuchi, H., Flooding water over a crowded urban district. *Journal of JSCE*, **491**(II-27), pp. 51–60, 1994. (In Japanese.)
[7] Uemura, Y. & Fukuoka, S., On the backslope stability of levee during the flood falling stage based on the levee vulnerability index. *Advances in River Engineering*, **24**, pp. 571–576, 2018. (In Japanese.)
[8] Muskat, M., *The Flow of Homogeneous Fluid through Porous Media*, McGraw-Hill, pp. 316, 1937.

TURBULENT KINETIC ENERGY IN FISH PASSES OF VARIOUS TYPES OF CONSTRUCTION

MARTA PUZDROWSKA & TOMASZ HEESE
Environmental and Geodetic Sciences, Faculty of Civil Engineering,
Koszalin University of Technology, Poland

ABSTRACT

A fish pass remains a hydraulic facility, which is to restore the migration possibility for ichthyofauna through the existing cross hydraulic structures. Except for the values of average flow velocities, the main features of flow deciding about the efficiency level of the device are values and characteristics of turbulent kinetic energy (TKE) distribution. TKE affects the power cost to be borne by fish to pass through the facility, and it affects the pace (i.e. the time) of passing through an obstacle. The main objective of research was to determine the impact of modifications in geometry of partitions in the bolt fishway channel on changes to hydraulic conditions and parameters for a water stream in a discharge channel. Relevant geometry of partitions directly improves the efficiency of objects. Bolt fishway is an object used for fish migration, where cross-partitions separating following chambers (pools) are made of cylindrical, rotary spandrel-beam elements of various height. The measurements were done for three instantaneous flow velocity components in indicated measurements sections. The analysis of the results was done for three-dimensional flow structure. The results were developed using Matlab software. The results were significantly different in comparison to the previously published ones. The differences referred to the values, as well as to the features of spatial distribution. This paper focuses on the analysis of TKE in bolt fishways, and on the comparison of that value's distribution in technical facilities of various types of construction, i.e. in pool-type fishways and in vertical slot fishways.

Keywords: fish pass, blot fishway, design of fish passes, turbulent kinetic energy, turbulence of water flow.

1 INTRODUCTION

Construction of efficient fish passes remains a very difficult and important task. A necessity of developing facilities of that type at water barrages results from the Water Framework Directive and from the Habitat Directive binding for the member states of the European Union. Those documents significantly consider improvement of quality for statuses and potentials of rivers and water reservoirs. However, construction of a fish pass without relevant control over its efficiency does not guarantee that quality rates for the environment would be improved. Regardless of the provisions valid in the EU, restoration of old fish migration routes is a global trend. Fragmentation of river habitats by hydraulic structures effects in decay of migrating species and, simultaneously, in high cost of environmental compensation through e.g. fish stocking.

The most popular type of facilities used for migration is a technical fish pass. Vertical slot fishways or pool-type fishways are the most often constructed ones. They are significantly easier to develop due to greater technical knowledge on the hydraulic conditions in those facilities. Technical fish passes may be developed easier on both the designing stage, as well as the investment's development stage. It affects the pace the object is being developed.

Based upon site observations it was stated that vertical slot fishways and pool-type fishways not always achieve the proper efficiency. The issue is especially related to fish species considered as so-called weak swimmers, and those are potamodromous fish. It was identified that they often wander around chambers (pools) of a fish pass or they cannot pass through the partition, thus they gather directly underneath the slot, inside the facility [1]. It

WIT Transactions on Ecology and the Environment, Vol 234, © 2019 WIT Press
www.witpress.com, ISSN 1743-3541 (on-line)
doi:10.2495/RBM190091

not only results in limitation of the possibility to migrate for weak swimmers, but what is most important, it significantly increases the migration time. It may result in fish reaching the spawning site too late.

Due to small efficiency of passing through fish passes by potamodromous fish, in case of previously known fish passes, research was undertaken to examine a different partition geometry for the technical facility. The examined type of fish pass is a bolt fishway. The obtained results were compared with data taken from publications related to facilities of another structure type. It is to establish the reasons for weaker migration of potamodromous fish. However, it also is to establish mechanisms for distribution of the most important – except for velocity – migration barrier, which is a value of turbulent kinetic energy (TKE).

2 MODELS AND METHODS

2.1 Description of the models

The name of the structure (bolt fishway) is associated with the applied cross-barriers in the facility's channel. Cross-partitions are made of linearly placed cylindrical bolt elements of various height, thus separating consecutive chambers (pools) of the fish pass. The cylinders are assembled on spindles and they may be rotated.

Cross-section of bolt elements – getting narrower with height – allows for forming diverse flow conditions in the fish pass just on the object's use stage.

The research was done based upon 2 laboratory models – A and B. A prototype of the solution is model A, which reflects basic assumptions for the construction. Model B remains a modification of model A, which includes moving of one bolt, and therefore moving the axis of partition for the main slot and for two smaller spillway slots.

Basic assumptions on bolt fish pass's geometry (in the actual scale) are as follows: channel width 4 m, length of a single chamber (pool) 6m, long drop of the bottom 2%. The designed actual flow was 2.14 m^3/s, at maximum filling of the chambers to a height of 1.10 m. Width of the main migration slot is 0.8 m. On the other hand, bolt elements have the following dimensions: no. 1 – height 1.3 m; no. 2 – height 0.6 m; no. 3 – height 0.8 m. The cross-section of all elements is elliptic and has the following diameter 0.64 m (semi-major axis) and 0.52 m (semi-minor axis). Bolt elements are assembled on steel spindles, and this is why the bolt element may be rotated and therefore the hydraulic conditions in the facility may be regulated on the object's use stage.

Models A and B were developed in a linear scale of 1:5, for which the remaining parameters are as follows: width 0.8 m (inside), and length of a single chamber 1.20 m (from the axis to the axis of partition). The model was developed as a series of eight chambers. Drop of the bottom was 2.0%. Maximum filling of the model's chambers was 22 cm, and minimum one was 20 cm. Bolt elements had the following parameters (numbering of bolt elements in accordance with Fig. 1): no. 1 – height 26 cm; no. 2 – height 12 cm; no. 3 – height 16 cm. The cross dimension was reduced to 12.8cm for the longer diameter, and to 10.4 cm for the shorter diameter. The measuring chamber was placed in a central part of the model. The test flow corresponding with the flow probability criterion according to Froude was 38 l/s.

For the purpose of tests three-dimensional acoustic Doppler velocimeter SonTek 16 MHz MicroADV was applied. Eleven sections were developed during the research, 18 measurement vertical lines each. Due to the fact that a distance between the measurement cell and the probe's head is 5 cm, there were huge difficulties in measuring final three spots placed at a level of 0 cm, 2 cm, and 4 cm underneath the surface of water.

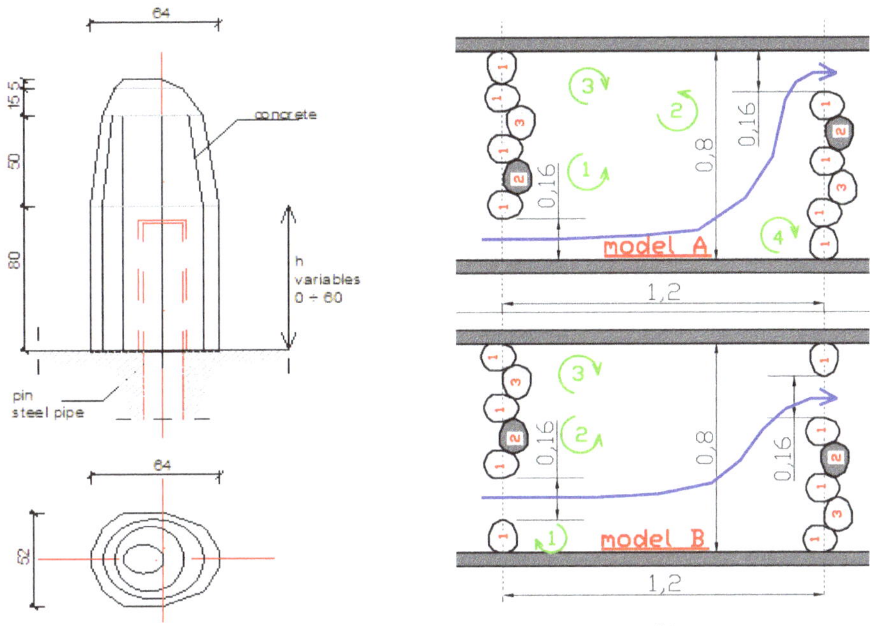

(a) (b)

Figure 1: Laboratory models. (a) Bolt element (cm); (b) Model A and B construction arrangement – dimensions of bolt elements quoted below. Note: location and direction of rotation for huge whirl structures is marked in green – marked with numbers from 1 to 4.

Due to requirements of Matlab, which was used to develop the results, one has decided to additionally measure those spots. Spots on a level of 4 cm underneath the surface of water (16 cm above the bottom) were measured by placement of the probe at an angle. It was enabled by a rotary handle fixing the probe. Then the flow was increased and measurement of the same spot was repeated, but this time at the vertical alignment of the probe. After analyzing the results it was established that the difference in values of average directional velocities and in their standard deviations is not more than 3% of the value. Therefore, it was assumed during the measurements that geometry of the bolt affects the flow parameters more than the distance between the measurement spot and the surface of water. Spots placed on a depth of 0 cm and 2 cm below the water table at filling of up to 22 cm were additionally measured at the increased flow.

It is important that the increase of flow (up to a height of 25 cm) did not cause flooding of the bolt chamber (height of bolt elements is 26 cm). This is why the stream's flow character in the model was kept. Active opening in the bolt's section was not modified at the additionally measured spots, because those levels were placed above the highest parapet of additional spillway slots.

The research was done at sections with the following spacing (18 measurement vertical lines at each section): section r – 0.00 m (bolt at the inlet to the measurement chamber); section a – 0.15 m; section b – 0.25 m; section bb – 0.35 m; section c – 0.5 m; section cc – 0.65 m; section d – 0.85 m; section dd – 1.00 m; section ddd – 1.05 m; section ee – 1.10 m; section rr – 1.20 m.

2.2 Methods of analysis

TKE (sometimes marked as “k”) remains one of the most important characteristics for the turbulent flow. As proved by the observations, its value highly affects behavior of fish in fish passes. From the physical point of view it is a kinetic energy of a liquid’s volume unit associated with pulsatory motion [2]. Therefore, a mathematic formula for TKE is as follows [3]:

$$TKE = 0.5\left[\overline{v_x'^2} + \overline{v_y'^2} + \overline{v_z'^2}\right], \tag{1}$$

where $\overline{v_x'^2}$ is the standard deviation for fluctuation of directional flow velocity.

The TKE grossing-up factor in the model amounted to 5.08. The research was developed using Matlab software.

3 RESULTS AND DISCUSSION

3.1 Results for bolt fishways

The research presented in this paper presents horizontal (parallel to the bottom) projection of flow parameters. In both cases A and B, the sections are placed just below the parapet of the biggest additional slot. They are placed on level no. 5, i.e. 10–11 cm over the model’s bottom (50–55 cm above the bottom, considering the actual scale). The results shown in Figs 2 and 3 present the outcome for the actual scale of the object (for Q = 2.14 m^3/s).

TKE distribution presented in Figs 2 and 3 show that the area of the main flow corresponds with quite low TKE value. Over the entire length of the main stream’s flow area TKE does not exceed the value of 0.1 m^2/s^2, for both of the models. The increase of TKE values was noted especially in whirl zones, at the boundary between the top flow area and the whirl zones. Induction of energy production in those areas remains a results of mutual impact of whirl structures on each other. The impact mainly results from a specific three-dimensional flow structure, which is highly different in each of the discussed cases.

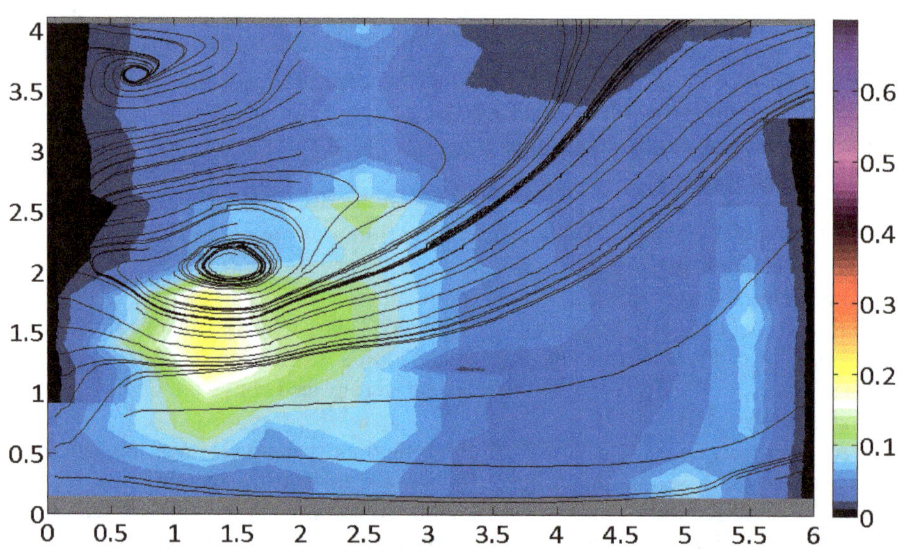

Figure 2: Characteristic section for TKE (m^2/s^2) – model A [4].

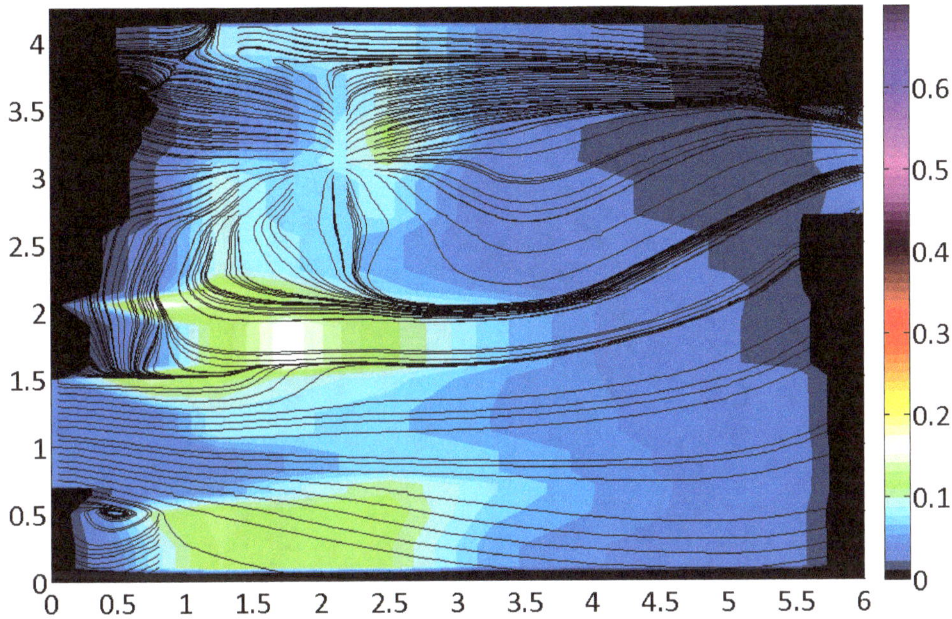

Figure 3: Characteristic section for TKE (m^2/s^2) – model B [4].

A detailed description of the spatial flow structure, TKE features, and changes in the distribution mechanism for that parameter shall be described in another paper.

3.2 Results for fish passes on the technical design – confrontation

In order to verify features of the flow in the bolt fishway – in reference to other types of the structure – we have analyzed several publications [1], [4]–[14]. Results of TKE tests obtained for the bolt structure are diametrically different from the ones previously known from the publications developed for other types of fish passes.

The first and the most important difference is the mechanism of spatial TKE distribution. Publications referring to technical fishways with pool-type and slot structure prove that the greatest TKE values in the chamber are focused around the outlet from the slot or from the migration hole.

One may generally see that the level values are taken by TKE in the area of the main water stream's flow. The whirl area contains areas, where TKE production is being noticeably reduced.

Domination of extreme TKE values in the area of the main flow remains a feature of both types of structures known before, i.e. pool-type structure and slot structure.

The other difference is the volume of obtained parameter. A direct comparison of TKE values using distributions given in various scientific articles is very difficult. There are two reasons for that: diversified scale of values for presented objects and normalization of TKE values applied by scientists.

Objects of major scale were compared to our model in the actual scale. Laboratory models remained a greater challenge, as they included much diversified flow values. At inclusion of

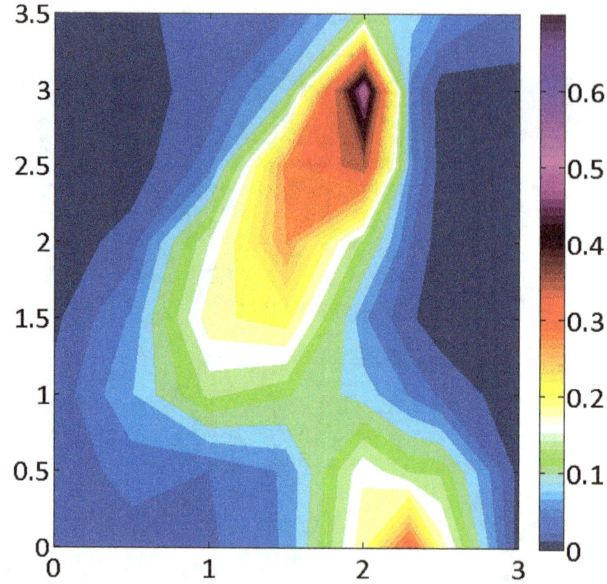

Figure 4: TKE distribution in a vertical slot fishway (m^2/s^2) (based on [5]).

variability of geometry features (e.g. in pool-type fishways) hydraulic conditions were often not corresponding with each other at all. The biggest issue was a huge difference of scale for filling of the facilities, and especially in case of pool-type structures. Therefore, at the analysis of TKE values, we decided to recalculate our model to a model comparable in terms of hydraulic parameters for a similar flow volume in the fish pass. This is why, for the purpose of rough comparison of our results with data taken from the publications, we projected the results of other research units, and – using Matlab software – recalculated the normalized TKE values to form defined in eqn (1).

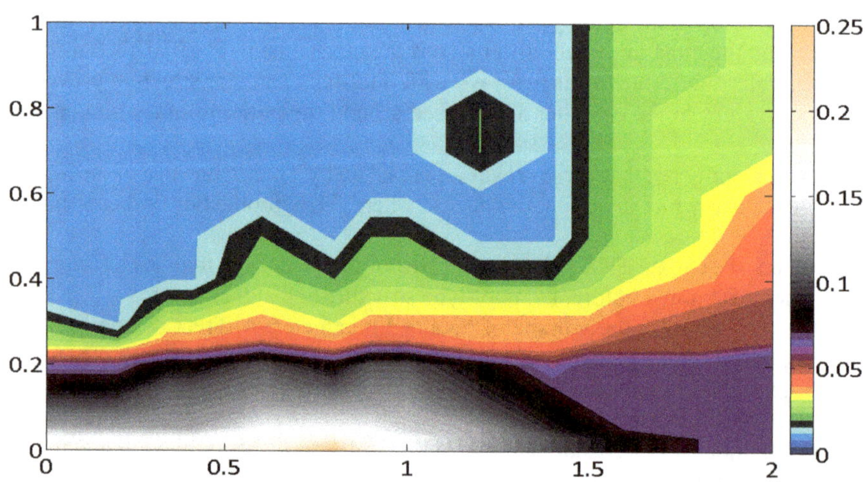

Figure 5: Distribution of TKE in pool-type fishway (m^2/s^2) [6].

This complicated and time-consuming task provided very interesting effects. It came out that the bolt fishway has a completely different distribution of TKE values in chambers. Furthermore, pool-type structures and slot structures – made in a similar scale – had significantly higher TKE values. The difference was noticeable in the area of the main stream's flow mainly.

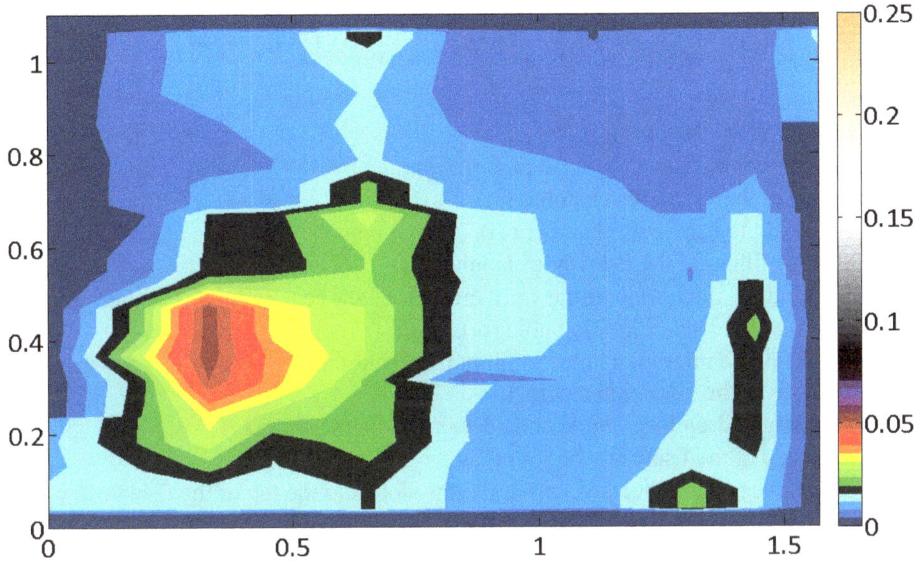

Figure 6: Distribution of TKE in bolt fishway – model A (m^2/s^2). Scale of the model according to [6].

3.3 Discussion

A significant stage of our works on that issue was establishing reasons for such a huge differences in TKE values in compared structures. As a result of the analyses a reason for a different behavior of the main stream was identified. The main issue was a diversified potential of the flow stream [4]. It turns out that bolt fishways have a stream with much bigger potential than the one recorded in fishways of another type.

We applied those assumptions to provide a series of simulations to verify how is the main stream's potential reflected in TKE values obtained just downstream of the outlet from the main migration slot in the facility. The analyses done allowed us to establish an approximate formula for $\sqrt{TKE}$ value. A basis of the formula is a relation determined in a formula (M). It was marked in orange on the diagram. The final formula (MH) was established through attempts to reach the expected value by the obtained parameter (M), i.e. $\sqrt{TKE}$. The most important achievement is the fact that the formula allows for forecasting of the TKE value in the central part of the outflow (just downstream of the slot) of the main stream in the fishway. It is feasible at application of data at the designers' disposal, without performance of laboratory tests. Those mainly are data related to the facility's geometry and basic hydraulic information available just at the preparation stage for the project. Examples used in the analyses are developed using various scales and various types of object structure. The formula reduces the issue of the object's scale. Some deviations of the estimated TKE value are obviously noticeable. However, one shall take into account the fact that the analyzed research was done by scientists in measurement spots located at various levels in reference to the fishway's bottom.

$$M = \left(\frac{1}{n}\right)\left[\left(\frac{Q}{L\cdot B\cdot h}\right)^2\left(\frac{l_o}{b_o\cdot h_o}\right)h^2\Delta H\right]^{0,5}\left(\frac{h}{\Delta H}\right)\left(\frac{h}{h_o}\right), \tag{2}$$

$$MH = \left(\frac{1}{n}\right)\left[\left(\frac{Q}{L\cdot B\cdot h}\right)^2\left(\frac{l_o}{b_o\cdot h_o}\right)h^2\Delta H\right]^{0,5}\left(\frac{h}{\Delta H}\right)^{1,1}\left(\frac{h}{h_o}\right)^{1,5}\left(\frac{h_{ou}}{h_o}\right)^4, \tag{3}$$

where

- n – number of migration slots (-).
- Q – consumption / flow in the fish pass (m^3/s).
- L – length of the fishway's chamber (m).
- B – width of the fishway's chamber (m).
- h – average filling of the fishway's chamber (m).
- l_o – longitudinal dimension of the slot / migration hole (e.g. thickness of the wall) (m).
- b_o – width of the slot / migration hole (m).
- h_o – height of the migration hole / minimum height of the slot's flooding downstream of the outlet from the migration slot (m).
- ΔH – hydraulic drop for a single chamber (m). Calculated as: $\Delta H = i[\%]\cdot L$.
- i – long drop of the fishway's bottom (%).
- h_{ou} – effective water gauge level affecting the slot from the top of the fishway's channel (m). Calculated as: $h_{ou} = h_o + \Delta H$. For the hole of the pool-type fishway $h_{ou} = h_o$.

The purpose of the test done was to achieve the value of √TKE by MH function. As it can be seen on Fig. 7, there is no full covering, when the MH function gets close to the value of √TKE. Noticeable deviations may however be justified. The most important difference between the measured TKE value and the value resulting from the formula is visible for case of publications with black and white diagrams were printed, what made the task more difficult [13]. Furthermore, in some case one may notice that the function provides slightly higher TKE value than the ones resulting from a direct measurement [7]. The issue may also refer to the proper location of measuring spots, and – to be honest – to not placing the probe in the area of higher TKE values in the model. In reality the flow in fishways is three-dimensional what proves how difficult this task is, and therefore the issue of proper placement of measuring spots remains a very complicated task.

4 CONCLUSIONS

Based upon the analyses done one may unequivocally state that the bolt fishway is a structure, which has the highest stream flow potential among the fish passes of technical type. It provides smaller – compared to other technical fishways – TKE values obtained in the area of the main flow at the inlet to the chamber. As a consequence, it is the most relevant facility for fish migration, including weak swimmers, i.e. potamodromous fish.

As reasons for diversification of the stream's potential within the fishway's channel were identified, a formula for an approximate TKE value – which may be expected in the central part of the slot just downstream of the stream's outflow from the migration slot – was developed. The formula allows for obtaining (approximate) information on the estimated TKE value on the stage of designing. It shall significantly facilitate the work and accelerate implementation of a potential investment.

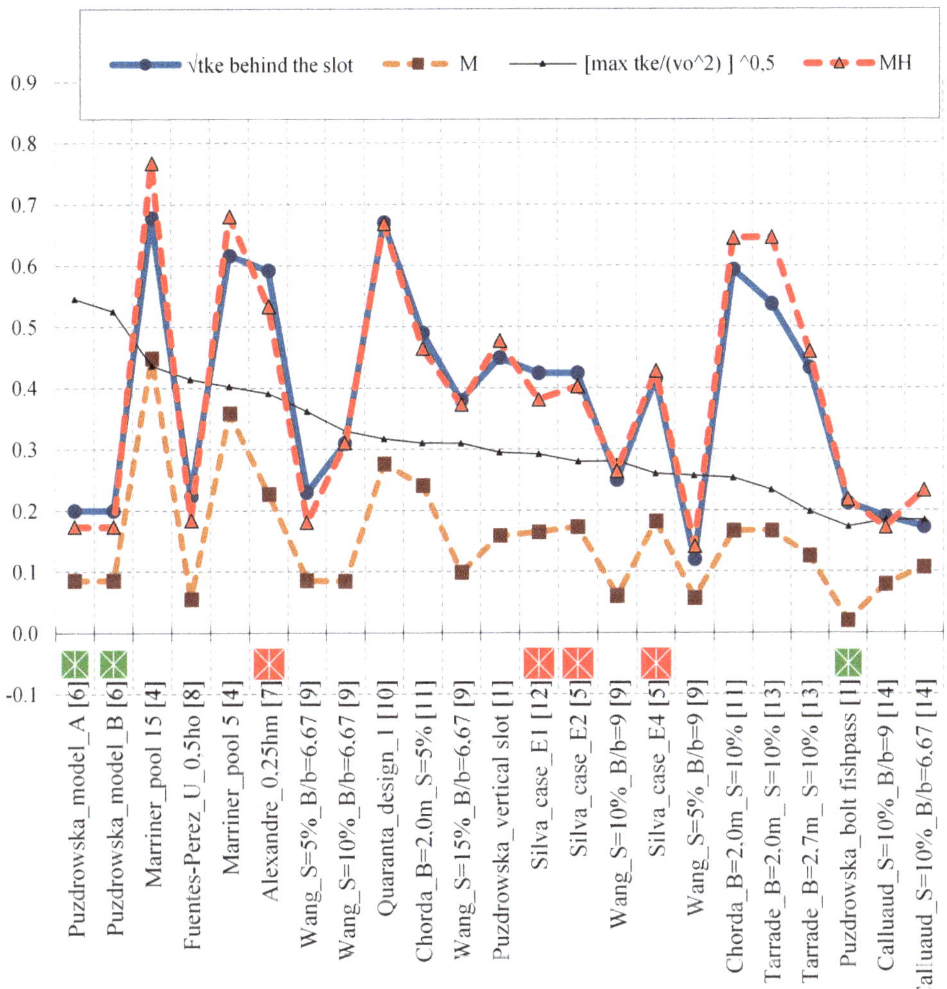

Figure 7: Diagram on the relation between the value of TKE and eqns (2) and (3).

REFERENCES

[1] Puzdrowska, M., Application of numerical methods in the design and analysis of fish pass efficiency. *Technical Transactions. Environment Engineering,* **110**(1-Ś), pp. 99–109, 2013.

[2] Gryboś, R., *Podstawy mechaniki płynów*, Wydawnictwo Naukowe PWN, Warsaw, Poland, 1998.

[3] Nezu, I. & Nakawaga, I., Turbulence in open-channel flows. IAHR Monograph Series, Balkema, Rotterdam, pp. 1–281, 1993.

[4] Puzdrowska, M. & Heese, T., Detailed research on the turbulent kinetic energy's distribution in fishways in reference to the bolt fishway. *Fluids*, **4**(2), p. 64, 2019. DOI: 10.3390/fluids4020064.

[5] Marriner, B.A., Baki, A.B.M., Zhu, D.Z., Cooke, S.J. & Katopodis, C., The hydraulics of a vertical slot fishway: A case study on the multi-species Vianney-Legendre fishway in Quebec, Canada. *Ecological Engineering,* **90**(2016), pp. 190–202, 2016.

[6] Silva, A.T., Katopodis, C., Santos, J.M., Ferreira, M.T. & Pinheiro, A.N., Cyprinid swimming behaviour in response to turbulent flow. *Ecological Engineering,* **44**(2012), pp. 314–328, 2012.

[7] Alexandre, C.M. et al., Use of electromyogram telemetry to assess the behavior of the Iberian barbell (*Luciobarbus bocagei* Steindachner, 1864) in a pool-type fishway. *Ecological Engineering,* **51**(2013), pp. 191–202, 2013.

[8] Fuentes-Pérez, J.F. et al., Hydraulics of vertical-slot fishways: Nonuniform profiles. *J. Hydraul. Eng.,* **145**(2), 06018020, 2019.

[9] Wang, R.W., David, L. & Larinier, M., Contribution of experimental fluid mechanics to the design of vertical slot fish passes. *Knowledge and Management of Aquatic Ecosystems,* **396**, 2010.

[10] Quaranta, E., Comoglio, C., Katopodis, C. & Revelli, R., Numerical simulations of flow field in vertical slot fishways. *XXXV Convegno Nazionale di Idraulica e Costruzioni Idrauliche,* Bologna, 14–16 September 2016.

[11] Chorda, J., Maubourguet, M.M., Roux, H., Larinier, M., Tarrade, L. & David, L., Two-dimensional free surface flow numerical model for vertical slot fishways. *Journal of Hydraulic Research*, **48**(2), pp. 141–151, 2010.

[12] Silva, A.T., Santos, J.M., Ferreira, M.T., Pinheiro, A.N. & Katopodis C., Passage efficiency of offset and straight orifices for upstream movements of Iberian barbel in a pool-type fishway. *River Research and Applications,* **28**, pp. 529–542, 2012. DOI: 10.1002/rra.1465.

[13] Tarrade, L., Texier, A., David, L. & Larinier, M., Topologies and measurements of turbulent flow in vertical slot fishways. *Hydrobiologia*, **609**, pp. 177–188, 2008.

[14] Calluaud, D., Pineau, G., Texier, A. & David, L., Estimation of the turbulent features of flow in vertical slot fishway: improvements on fishway design criteria. *3rd IAHR Europe Congress, Book of Proceedings*, Porto, Portugal, 2014.

CRITICAL ROLE OF BRIDGING ORGANIZATIONS IN RIVER BASIN MANAGEMENT IN ALBERTA, CANADA

JUDY STEWART
Canadian Institute of Resources Law, University of Calgary, Canada

ABSTRACT

The paper examines bridging and brokerage functions performed by a Canadian watershed-scale bridging organization in Alberta's watershed governance and management system. The Bow River Basin Council (BRBC), a multi-stakeholder bridging organization in the Bow River Basin in southern Alberta, Canada, provides the demonstration context for exploring concepts of bridging organizations and their evolving roles. BRBC performs strategic bridging functions that connect and engage cross-sectoral public and private stakeholders who otherwise would not be included in watershed governance or management decision-making processes. BRBC operates at the watershed-scale and plays critical roles as a strategic broker of information, knowledge, values, and power and influence in the region. BRBC provides venues for stakeholder collaboration to resolve complex watershed management problems where solutions are reached by consensus. Stakeholders work together to identify shared community values and issues of common concern. BRBC builds trust relationships, co-generates cross-sectoral knowledge, and facilitates social learning to help resolve stakeholder conflicts over the use and management of scarce water resources in the Bow River Basin.

Keywords: bridging organization, social-ecological systems, social network analysis, social network mapping, strategic bridging functions, social learning, river basin management.

1 INTRODUCTION

River basins are complex, dynamic social-ecological systems (SES) where society and the ecosystem are inextricably connected, co-evolving and co-regulating cross-scalar adaptions [1]. In the context of river basin governance and management in Alberta, Canada, bridging organizations are emergent social-political arrangements that address system dynamics and complexity [2]. Alberta's population continues to grow, notwithstanding recent economic constraints associated with the downturn in the oil and gas industry [3]. Alberta's social and cultural norms continue to adapt to river basin management science that predicts future water scarcity and deteriorating water quality [4]. New standards, codes of practice and regulations for land use development are being imposed by government to keep the state of the environment within Alberta's seven major river basins [5] within desirable bounds [6].

However, ecosystems do not respect human-made geo-political boundaries or cultural norms and practices [7]. River basins, such as the Bow River Basin (the Bow Basin), a tributary to the South Saskatchewan River in southern Alberta continue to evolve and adapt to human interventions and constraints. Dams, weirs and shoreline modifications have been imposed on riverine systems and adjacent landscapes to accommodate human settlement and economic development. Floods and droughts are both unpredictable and imminent due to climate change and human-induced landscape changes throughout the basin [8].

Governing and managing human interactions and use of the land and water in the Bow Basin involves complex and evolving policy and regulatory systems. In 1992, the Bow River Basin Council (BRBC) was appointed by Alberta's Minister of the Environment to evaluate the state of the Bow Basin and provide advice to the Government of Alberta (GOA) on how to address deteriorating water quality south of the City of Calgary's wastewater treatment plant [9]. Originally, BRBC members were appointed by Ministerial Order [9]. Over time, the BRBC emerged as a self-organizing and self-regulating society comprised of self-

WIT Transactions on Ecology and the Environment, Vol 234,
www.witpress.com, ISSN 1743-3541 (on-line)
doi:10.2495/RBM190101

selecting volunteer representatives from the GOA, municipalities, industry, non-government organizations, academia, and the public [9]. BRBC stakeholders are self-interested with competing interests in water and landscape management, but they use consensus decision-making processes to address identified common concerns about water quality, water quantity and river basin resilience [9].

The purpose of this paper is to explore the critical role of bridging organizations in river basin management in Alberta using BRBC as the demonstration context. First, Alberta's complex river basin governance and management system is briefly reviewed in the context of legal pluralism. Second, BRBC's social network structure is presented along with the critical bridging and brokering functions BRBC is able to perform. The paper concludes that BRBC is structured and functions as a bridging organization, connecting stakeholders who would otherwise not be connected to solve complex river basin management problems. BRBC critical strategic bridging functions have influenced increased municipal participation in several river basin management activities in the basin [2], [10].

2 ALBERTA'S COMPLEX RIVER BASIN MANAGEMENT SYSTEM

In the Bow Basin (Fig. 1) where BRBC operates, legal pluralism dominates the complex system for river basin governance and management [10], [11]. Legal pluralism focuses on a multitude of legal orders within one social field, for example how different sectors of society divert and use water in a semi-arid region [11]. Teubner [12] referred to legal pluralism as being "social norms and legal rules, law and society, formal and informal, rule-oriented and spontaneous". Therefore, in situations where legal pluralism dominates, conflict may arise when implementing rules that seem to conflict with established cultural norms and social customs [11]–[13].

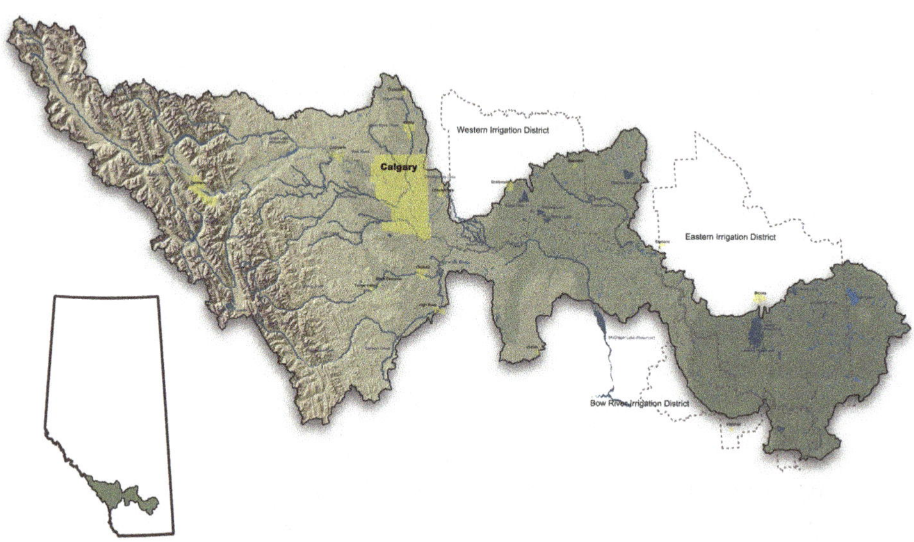

Figure 1: Map of the Bow Basin, Alberta, Canada [9].

Historically, Alberta inherited the British common law system of riparian rights to water, making settlement of the semi-arid region difficult because only those who owned land adjacent to water bodies could divert and use surface water. Settlers could also pump groundwater from under the lands they owned to the surface and could store and use it pursuant to the British common-law rule of capture. Over time, as governments and self-interested corporations across the country encouraged settlement of Alberta's landscape, the GOA claimed ownership and the right to divert and use all surface and groundwater in the province. Water law provisions were enacted to allow the development of complex irrigation reservoirs, weirs and canals to transport water from rivers such as the Bow to dry lands to encourage settlement and agricultural operations. Alberta's first-in-time-first-in-right system of water allocation licenses emerged. Irrigation districts and the City of Calgary were issued large water allocation licenses with no expiry date [1], [9].

With this complex system of water allocation licensing, water approvals to disturb water bodies, environmental laws, and laws to create and regulate irrigation districts and local governments, there are no shortage of provincial policies and laws that affect how water and land are managed in the basin [10], [11]. However, historically, there has been little integration of these multi-jurisdictional, overlapping and sometimes conflicting policies, laws, regulations, codes of practice, guidelines, operating protocols, negotiated rules, municipal statutory planning documents, municipal bylaws, and BRBC's watershed or river basin management plan (RBMP) [10], [11].

Within the GOA, water, air, public and private lands, biodiversity, energy, public health and the economy are all regulated through departmental silos and substantive laws [10], [11]. Several different policies, laws and regulations regulate human diversion and use of water and land use development, depending on whether the land is privately or publicly owned or belongs to the Government of Canada (GOC) [10], [11]. While both water and land are managed as distinct resources or components of the environment [14] there is little integration within the complex system of policy and law. While Water Act [5] provisions provide for the development of water management plans through collaboration with partners, only "approved water management plans" must be considered by the Director when approving water licenses or approvals. There are only two approved water management plans in the province, and one of them exists for the South Saskatchewan River Basin in which the Bow is a significant tributary [15]. Water management plans and watershed or RBMPs are not the same, nor are they integrated in any of Alberta's seven major river basins.

RBMPs, such as BRBC's co-created plan [16] emerge through voluntary stakeholder collaboration. Generally, they reflect GOA's partnership objectives as stated in Water For Life: Alberta's Strategy for Sustainability (Water For Life) [17]. RBMPs are not legal instruments that are enforceable by the courts, although they often recognize and reflect the legal system that exists at the time they are created. Several dozen policies, laws and other legal instruments [10], [11] address how people interact with water and land, but how people interact in river basins is not regulated.

Alberta's current legal regime for both water and land use governance and management has evolved over time as legislatures adapted to emerging social and ecological phenomena [10], [11]. For example, the Municipal Government Act [18] has been amended several times since the mid-1990s with many significant amendments enacted between 2015 and 2018 that affect how municipalities may address local and transboundary environmental matters [18], [19].

The GOA has delegated to municipalities the authority to create policies and enact bylaws to regulate and control environmental matters within their boundaries, as long as the local policies and bylaws do not conflict with provincial or federal enactments [18], [19]. Every municipality in the Bow Basin has enacted municipal land use plans and land use bylaws. In addition, some municipalities have water conservation policies and bylaws, as well as water distribution and pricing bylaws in place. Some have enacted bylaws to prohibit, or regulate and control inappropriate land uses that may negatively impact water resources within their boundaries [10]. All of the municipalities in the Bow Basin voluntarily participate in several water resource management activities [10].

In addition to provincial and municipal policies, laws, regulations, bylaws, codes of practice, guidelines and best practices, federal laws enacted by the GOC supersede all other legal instruments that address similar subject matter. For example, there are federal laws that regulate boating, inland fisheries, habitat for species at risk and migratory birds, and environmental impact assessments on federal lands [10]. Nonetheless, regulatory gaps between federal, provincial and municipal laws persist at the river basin scale where no one level of government has sole jurisdiction [9]–[11].

In a situation of legal pluralism like this, conflicting demands, different scales, and cultural orientations may cause uncertainty for individuals who cannot be sure which rules will apply in a specific situation [13]. Self-interested groups and individuals may forum-shop to select rules that advance their personal or corporate objectives, posing challenges to different legal authorities who all claim jurisdiction over the subject matter [13], for example development of wetlands or riparian corridors, or along shorelines and flood risk areas that provide critical habitat for fish, migratory birds and endangered species [19].

Luhmann [20] suggested that any legal system is a self-regulating subsystem of society that functions to stabilize normative expectations of all the other social subsystems, for example, politics, science and economics. Practically speaking, the legal subsystem of any society defines what comprises the law [20]. In this context, RBMPs that are co-created by multi-stakeholder volunteer organizations such as BRBC are not law. The organizations have no delegated authority to make rules or decisions about water allocation, to enforce rules preventing water quality degradation, or to resolve conflicts among competing stakeholders. For example, BRBC cannot require that stakeholders implement the strategies in the RBMP, nor can the organization impose government sanctions for non-compliance with plan objectives or policies [2], [9], [10].

Stewart [10], who interviewed the Board of Directors of BRBC in 2016, determined that stakeholders in BRBC recognize the complex, dynamic legal system at play in the river basin, and that it is enforced by GOA departments, municipal governments and the court system [9]. Through the RBMP, BRBC provides advice to the GOA about emergent river basin management issues and proposes management objectives and strategies for implementation by its own stakeholders. In addition, through its Board of Directors, and standing committees for policy and technical advancements, BRBC identifies and attempts to address regulatory gaps in provincial and municipal policies and laws through recommendations to government bodies in both formal and informal submissions [9], [10].

BRBC operates at the nexus of Alberta's overlapping social systems, for example law, politics, social programs and economics, and bridges what is the law with what is not the law [2], [10]. As a cross-sectoral, multi-stakeholder network with representatives from both government and civil society, BRBC operates in oscillating spaces where legal pluralism dominates, but where no one law is definitive [10].

3 BRBC OPERATES IN COMPLIANCE WITH ALBERTA'S LEGAL SYSTEM

One of the GOA's strategies to achieve water governance and management outcomes provided in Water For Life [17] is to develop partnerships between all levels of government and non-government actors at a river basin scale. A river basin in this context is not limited to the seven major river basins in the province, but is considered the area of land that catches precipitation and drains into a larger body of water such as a marsh, stream, river or lake [17]. Several multi-stakeholder groups, called watershed advisory and planning councils (WPACs) and watershed stewardship groups (WSGs) exist in Alberta and operate at different landscape scales pursuant to Water For Life. BRBC was Alberta's first WPAC and set the stage for all others in the province [9]. Several WSGs operate along the tributaries of main stem rivers, such as the Bow River, and many are members of BRBC.

However, while the GOA provided some guidance to WPACS and WSGs about preparing RBMPs, the GOA did not require or recommend that WPACS or WSGs adopt or implement Water For Life objectives or strategies when developing their plans. While BRBC is sanctioned through Water For Life to advise the GOA about river basin governance and management issues, and to create state of the basin reports and the RBMP, the organization is not compelled by law to perform any of its functions in accordance with the water governance and management legal system [17].

While generally in compliance with Alberta's complex legal system for water and land management, BRBC's most critical function in river basin governance and management is as a bridging organization, connecting stakeholders who would otherwise not be connected [2], [21]. Stewart [10] discovered that during BRBC's meetings and functions, provincial and municipal decision-makers are connected and collaborate with the stakeholders who benefit from licensed water use, approved levels of substance releases, and land use development and subdivision approvals. Through BRBC's strategic bridging processes [7], [22], government officials also collaborate with and develop trusting relationships with other stakeholders who are affected by those same licensed water uses or land use approvals [10].

4 BRBC IS STRUCTURED AND FUNCTIONS AS A BRIDGING ORGANIZATION

The concepts of bridging organizations and strategic bridging functions are not new [10]. In the context of environmental governance, Crona and Parker [7] summarized that a bridging organization is one that links diverse actors or groups with stakes in resource management through some form of strategic bridging process. Bridging organizations are distinct from the organizations or stakeholders they want to connect. Historically, bridging organizations emerged in the social arena in developing countries to purposefully connect disadvantaged people with resource providers and other service organizations because, otherwise those in need were not aware of those who could help them [22]. In the social arena, bridging organizations function around the globe at all scales, from local to multi-national [22].

The degree of organizational formalization and interpenetration of members within the groups makes bridging organizations different from roundtables and task forces and other informal multi-stakeholder groupings that get together to address specific resource management problems. Table 1 below illustrates the criteria that define a bridging organization according to Crona and Parker [7].

Reid [22] identified nine distinct strategic bridging functions, as follows: to maintain the status quo of an arrangement; for problem solving and transformation; to facilitate collective action; to facilitate capacity building; to increase impact or autonomy; to gain legitimacy or resources; to mediate norms among actors; to develop compromises; and to support activist voices in negotiations with dominant actors [22]. Bridging organizations have also been found necessary for social learning in transdisciplinary settings, where social learning is "learning that occurs when people engage one another, sharing diverse perspectives and experiences to develop a common framework of understanding and basis for joint action" [23].

In the context of watershed resiliency, Walker and Salt [24] summarized that bridging organizations play critical roles in natural resource management systems because "they build local institutions; horizontal linkages; vertical linkages; and increase public education and

Table 1: Criteria that define a bridging organization [2], [7], [10].

Organizational Structure	Membership	Primary Objective	Goals
Formalized organization with own resources and personnel.	Varying levels of stakeholder diversity with high degree of actor interpenetration.	Develop a strategic bridging process to connect otherwise unconnected actors to a network. Third party to those it seeks to connect.	1. Provide an arena for: • Learning • Co-creation of knowledge • Building trust • Conflict resolution. 2. Act as facilitators, mediators, and negotiators. 3. Attract expertise, knowledge and resources.

innovations. They mediate connections between otherwise unconnected actors, attract new knowledge and resources from outside the natural resource management system required for social learning and transformation, and, in doing so, increase the system's adaptability and resilience".

As explained by Crona and Parker [7], not all multi-stakeholder organizations have appropriate governance or social network structures among actors to function as bridging organizations. Stewart [10] used social network theory [25] and conducted hour long interviews with the BRBC Board of Directors and eighteen municipal representatives in the Bow Basin to determine BRBC's social network structure and approach to cross-sectoral, transdisciplinary and transboundary river basin governance. Stewart [10] created data sets from the interviews using letters and numbers to protect confidentiality of network actors. Using *Pajek* social network mapping software, that also analysed the different functions performed by different network actors, Stewart [10] translated the data sets into a social network map depicted in Fig. 2 below. As explained by de Nooy et al. [25] "social relations are considered channels that transport information, services, or goods between people and organizations. In this perspective, social structure helps explain how information, goods, or even attitudes and behavior diffuse within a social system. Network analysis reveals social structure and helps to trace the routes that goods and information may follow. Some social structures permit rapid diffusion of information, whereas others contain sections that are difficult to reach".

Stewart [10] observed that certain actors in the periphery of the BRBC's social network were critical bridges and brokers of information, knowledge, values, power, and influence in the wider polity. She discovered that these actors were only loosely tied to the BRBC network through other BRBC stakeholders who were all central figures with strong ties to the organization's core [10]. BRBC's particular social network structure made it possible for the organization to perform critical strategic bridging functions in river basin governance and management. BRBC was able to facilitate municipal collaboration and increased municipal participation in river basin management activities [10]. The bridging or weak ties in BRBC's periphery were critical for brokerage of information and diffusion of necessary innovations to transform how municipalities participated in river basin management activities [10]. Stewart determined that the same network structure that enabled BRBC to function as a

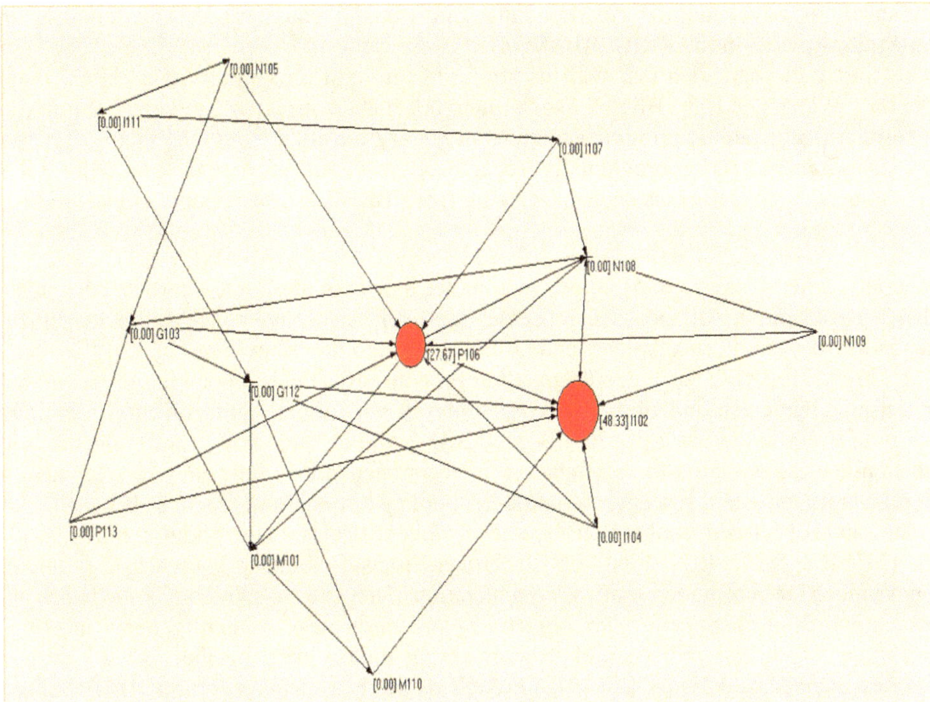

Figure 2: BRBC's social network using *Pajek* software [10], [25] illustrating the strong core and loosely connected actors in the network's periphery.

bridging organization also helped it achieve rapid diffusion of new information, best management practices and social learning throughout the Bow Basin [10].

BRBC did not deliberately set out to perform these strategic bridging functions [9], [10]. However, Stewart [10] verified through interviews that BRBC connected stakeholders who would otherwise not be connected; helped stakeholders identify shared resource management values and issues of concern; provided opportunities and venues for stakeholders with competing interests to build relationships of trust; collected information and transdisciplinary knowledge to create state of the basin reports and the RBMP; and provided venues and opportunities for social learning both within the organization and throughout the province [2].

Stewart [10] verified that BRBC has become a valued local river basin governance and management institution providing horizontal linkages between all levels of government, and vertical linkages between municipalities [9]. Through their quarterly information sharing forums they increase public education about emergent policy, technology and new innovations [9]. While bringing together previously unconnected representatives from different sectors, they attract new information and resources from outside the Bow River Basin. As social learning occurs, BRBC fills gaps in the complex regulatory system by showcasing strategies for improved management systems in the quarterly education forum, which eventually leads to transformations in industry practices [9]. As a result, in some cases, the RBMP's strategies have led to increased resiliency in the basin, and flexible and adaptive processes for achieving shared objectives described in the plan [9], [10], for example several municipalities have adopted riparian land and wetland conservation and management strategies as set out in the BRBC RBMP [6].

Stewart [10] concluded that with its strong history and recognition as a WPAC under Alberta's Water For Life, BRBC's social network has the structural integrity required to perform critical strategic bridging functions necessary for effective river basin governance and management. As a provincially recognized and funded organization with GOA stakeholders at the consensus decision-making table, BRBC is a sustainable organization.

5 BRBC'S CRITICAL ROLE IN RIVER BASIN MANAGEMENT

Generally, when cross-sectoral, multi-stakeholder networks of affected parties, such as the BRBC participate in collaborative processes for river basin governance and management, they build trust among stakeholders and influence stakeholder behaviours [2], [7], [10], [21]–[23], [26], [27]. BRBC has a good reputation for consensus-decision making processes and establishing agreement among self-interested parties who are all competing for the limited supply of water in the basin [9]. In this way, BRBC wields some power and influence with decision-makers as the go-to organization for recommendations for policy development to fill regulatory gaps, and strategies to implement policy through best practices [9], [10].

Stewart [10] determined that BRBC performs four critical strategic bridging functions [2], [7], [21], [22], [26], [27]. First, BRBC brokers transdisciplinary knowledge, providing opportunities for information sharing, co-generation of new transdisciplinary knowledge, and co-creation of management plan objectives and strategies. Annual general meetings, workshops, committee projects and programs bring stakeholders together to work through complex management issues [2], [9]. Through quarterly education forums, BRBC shares emergent information and knowledge about resource management technology, management planning and implementation, stewardship activities, and new provincial-scale policy and legislation [2], [9], [10]. BRBC is noted for the critical work done by standing committees for communications, policy development and technical advancement [9]. Actors in BRBC's network periphery share information generated through BRBC's collaborative processes with other WPACs and WSGs in other river basins. The GOA is not in the position to share information and co-create knowledge necessary for social learning and system transformation on its own.

Second, BRBC brokers shared river basin management values. The organization provides venues and resources to mediate between values, negotiate desired outcomes, create agreed upon management scenarios, and reach consensus on strategies to improve river basin management [2], [9], [10]. These values have emerged over time through connections made between stakeholders with competing interests who share common concerns about water scarcity or degradation and land use management adjacent to water bodies. BRBC's consensus decision-making processes help members to identify and prioritize cross-sectoral management values. The GOA's substantive regulatory system is unable to differentiate between and manage for values basin by basin, as laws and regulations necessarily apply right across the province as one-size-fits-all.

Third, BRBC brokers power and influence. BRBC influences provincial and municipal decision-makers in the river basin by providing venues and opportunities for interactions between stakeholders and government officials, and through presentations and written submissions. Stewart [10] identified that social learning that took place through BRBC collaborative processes was instrumental in the systemic transformation in how municipalities regulated and controlled human activities that may negatively impact land, water, and air quality. Without such transformative action, municipal land uses might have continued to fragment the structural and functional connectivity of the landscape at the nexus of water and land use [1], [2], [9], [10]. While the GOA has promulgated provincial land use policies for conserving and managing water and natural resources since the Municipal Government Act [18] was enacted in 1994, these policies were not being applied by municipalities with any consistency in the Bow Basin [18], [19].

Last, BRBC promotes voluntary collective action for the improvement of the health and resiliency of the river basin. Stakeholders who attend BRBC's quarterly forums are exposed to emergent policy and technical advancements and choose to implement best management practices when and where they can. The GOA cannot manage the basin by themselves through substantive regulatory processes, and the partnership with BRBC has proven beneficial to the health, sustainability and resiliency of the basin [9].

6 CONCLUSION

In this Canadian context, BRBC is structured and functions as a bridging organization, performing critical strategic bridging functions for effective governance and management of land and water in the Bow Basin. Through collaborative processes, provincial and municipal levels of government are connected with various stakeholders with competing interests in how water and land are to be used in times of rapid growth and water scarcity.

BRBC helps stakeholders identify shared interests and common values for river basin management. It also plays a critical role in social learning and policy development. Gaps in provincial and municipal policy and law are identified through BRBC governance processes and recommendations are made to the GOA for addressing those gaps. The co-creation of the state of the basin reports and the river basin management plan lead to new co-created transdisciplinary knowledge, best practices and changes in how people interact in the environment. Significantly, BRBC influences increased municipal participation in river basin management activities throughout the basin.

In Alberta, with its complex legal system, provincial and municipal governments understand the value of partnering in collaborative governance processes with BRBC and other WPACs around the province. This paper used BRBC's as the demonstration context to discuss the requisite social network structure for WPACs to perform several critical strategic bridging processes for effective watershed governance and management. Further research may be required to determine if Alberta's other WPACs have a similar social network structure and perform similar functions.

It may be that GOA may no longer be able to manage complex, dynamic human interactions at the river basin scale that impact the quality and quantity of scarce water resources through substantive regulatory regimes alone. They may require active participation of bridging organizations working alongside and enhancing the effectiveness of the regulatory system through voluntary collective action. The local institutions that have emerged in Alberta, such as BRBC and other WPACs, and the horizontal and vertical linkages between actors that have been made have helped to increase public awareness of the need to collaborate to solve complex river basin management problems. Solving these problems needs to be done at a basin-scale where shared community values can be used to establish trade-offs and develop triggers and thresholds for keeping the quantity and quality of water resources within desired bounds. Bridging organizations in Alberta's river basin governance and management system are critical to attracting knowledge and resources otherwise not available to governments, and for introducing adaptions for system transformation. Without system transformation, the resiliency of the Bow Basin may be put at risk during periods of sustained rapid growth in population and economic development.

REFERENCES

[1] Tyler, M.E. & Quinn, M., Identifying social-ecological couplings for regional sustainability in a rapidly urbanizing water-limited area of western Canada. *Wessex Sustainability Development and Planning VI*, WIT Press: Southampton and Boston, pp. 175–191, 2013.

[2] Stewart, J. & Tyler M.E., Bridging organizations and strategic bridging functions in environmental governance and management. *International Journal of Water Resources Development*, pp. 1–24, 2017.

[3] Government of Alberta, Economic outlook: Overview. www.alberta.ca/budget-economic-outlook.aspx#toc-9. Accessed on: 2 Jan. 2019.

[4] AMEC, *South Saskatchewan River Basin in Alberta: Water Supply Study,* Alberta Agriculture and Rural Development: Lethbridge, Alberta, 2009.

[5] Government of Alberta, Water Act, R.S.A. c.W-6. www.qp.alberta.ca/documents/Acts/w03.pdf. Accessed on: 20 Dec. 2018.

[6] Government of Alberta, South Saskatchewan Regional Plan 2014–2024. https://open.alberta.ca/dataset/460ac866-4416-4d77-a25a-a02fab85a6ec/resource/8261ce03-aa0f-4621-8e2d-c610a72ac37c/download/south-saskatchewan-regional-plan-2014-2024-february-2017.pdf. Accessed on: 15 Dec. 2018.

[7] Crona, B.I. & Parker, J.N., Learning in support of governance: Theories, methods, and a framework to assess how bridging organizations contribute to adaptive resource governance. *Ecology and Society,* **17**(1), 2012.

[8] Pomeroy, J.W., Stewart, R.E. & Whitfield, P.H., The 2013 flood event in the South Saskatchewan and Elk River basins: Causes, assessment and damages. *Canadian Water Resources Journal*, **41**(1–2), pp. 105–117, 2016.

[9] Stewart, J. & Bennett, M., Integrated watershed management in the Bow River basin, Alberta: Experiences, challenges, and lessons learned. *International Journal of Water Resources Development*, **33**(3), pp. 458–472, 2017.

[10] Stewart, J.A., reflexive legal framework for bridging organizations in regional environmental governance and management. PhD dissertation, University of Calgary, 2016.

[11] Stewart, J., Reflexive approaches to lawmaking to legitimize environmental bridging organizations in the Calgary region. *Journal of Environmental Law and Practice*, **31**(1), pp. 37–77, 2017.

[12] Teubner, G., Substantive and reflexive elements in modern law. *Law and Society Review*, **17**(2), pp. 239–286, 1983.

[13] Tamanaha, B.Z., Understanding legal pluralism: Past to present, local to global. *Sydney Law Review*, **30**(3), pp. 375–411, 2008.

[14] Environmental Protection and Enhancement Act, c. E-12. www.qp.alberta.ca/documents/acts/e12.pdf. Accessed on: 4 Jan. 2019.

[15] Government of Alberta, Approved Water Management Plan for the South Saskatchewan River Basin (Alberta). https://open.alberta.ca/publications/0778546209. Accessed on: 4 Jan. 2019.

[16] Bow River Basin Council, Bow Basin Watershed Management Plan. Phase 1: Water Quality. Bow Basin Watershed Management Plan Steering Committee, 2008. https://brbc.ab.ca/our-activities/bow-basin-watershed-management-plan. Accessed on: 4 Jan. 2019.

[17] Government of Alberta, Water For Life: Alberta's Strategy for Sustainability. https://open.alberta.ca/publications/0778530582. Accessed on: 4 Jan. 2019.

[18] Government of Alberta, Municipal Government Act, R.S.A. 2000, c.M-26 and regulations. www.qp.alberta.ca/documents/Acts/m26.pdf. Accessed on: 3 Jan. 2019.
[19] Stewart, J., Do recent amendments to Alberta's Municipal Government Act enable management of surface water resources and air quality. *Alberta Law Review*, **55**, pp. 1009–1044, 2017.
[20] Luhmann, N., *Law as a Social System*, Oxford University Press: New York, 2004.
[21] Rathwell, K.J. & Peterson, G.D., Connecting social networks with ecosystem services for watershed governance: A social-ecological network perspective highlights the critical role of bridging organizations. *Ecology and Society*, **17**(2), pp. 24–45, 2012.
[22] Reid, K.E., Building a framework for institutional change. PhD dissertation, University of Texas, Austin. p. 48, 2004.
[23] Schusler, T.M., Decker, D.J. & Pfeffer, M.J., Social learning for collaborative natural resource management. *Society and Natural Resources*, **16**(4), pp. 309–326, 2003.
[24] Walker, B. & Salt, D., *Resilience Thinking: Sustaining Ecosystems and People in a Changing World*, Island Press: Washington, DC, 2006.
[25] de Nooy, W., Mrvar, A. & Batagelj, V., *Exploratory Social Network Analysis with Pajek*, 2nd ed., Cambridge University Press: Cambridge, p. 139, 2011.
[26] Bodin, O. & Prell, C. (eds)., *Social Networks and Natural Resource Management: Uncovering the Social Fabric of Environmental Governance*, Cambridge University Press: New York, 2011.
[27] Berkes, F., Evolution of co-management: Role of knowledge generation, bridging organizations and social learning. *Journal of Environmental Management*, **90**, pp. 1692–1702, 2009.

ASSESSMENT OF CLIMATE CHANGE IMPACTS ON THE HYDROLOGIC RISK BY QUANTITATIVE ALTERATION IN THE CATALAN INTERNAL BASINS

LAURA RAMOS SOLER[1], JAVIER PAREDES ARQUIOLA[1], JOAQUÍN ANDREU ÁLVAREZ[1] & ANTONI MUNNÉ TORRAS[2]
[1]Universitat Politècnica de València, Spain
[2]Agencia Catalana del Agua, Spain

ABSTRACT

The Catalan Internal Basins has a peculiarity in respect of other Mediterranean basins; its total urban demand exceeds the agriculture extractions, so the returns of the wastewater treatment plants have special relevance in the total water balance. Usually, the main quantitative alterations of the hydraulic regime of the rivers are the regulation of reservoirs and the demands. The main objective of this study is to include and evaluate the quantitative pressure of urban discharges on the analysed system, in order to estimate the altered regime risk of water bodies and analyse their evolution in future scenarios of Climate Change. The tool used to achieve these goals is the Rapid Response Environmental Status (R^2EA) model. Firstly, from results, the percentage of the volume dumped over the total inflows of each of the bodies is calculated and the simulated period is evaluated. Secondly, the percentage of water extraction is calculated from discharges located upstream of the intake for each of the urban demands included in the model. These indices are calculated for three Climate Change scenarios. The calculated indices analysed show that the water regime alteration by discharges mainly affects temporal water bodies. In addition, a significant risk increase during summer months is identified. Finally, we propose that the water managers should use this approach in order to characterize the vulnerability level by quantitative alteration of aquatic ecosystems.

Keywords: water model, large scale, R^2EA, climate change, environmental risk.

1 INTRODUCTION

The direct and indirect consumption of natural resources is often accompanied by a degradation of the affected ecosystems. The main reasons are overexploitation and pollution [1]. Access to water resources is fundamental for the development of most human activities, as well as being essential to ensure a good level of quality of life for the populations [2].

Although it is a renewable resource due to the water cycle, in some regions the renewal capacity is lower than the rates of extraction or the pollution level, which is a challenge for the organizations responsible for the good condition of inland aquatic ecosystems [3], [4].

Increasing pressures on water resources increases the complexity of managing natural systems, especially in arid or semi-arid areas characterized by marked episodes of drought. The vulnerability of these zones creates the need to apply specific management techniques and strategies that promote the sustainable use of water [5], [6].

This situation has fostered the development of numerous indices that quantitatively reflect the use of this resource in a territory [7], [8]. Some examples are the Water Stress Index, Water Exploitation Index, Water Allocation or Exploitable Water Resources, among others.

In the present study, a specific index has been developed to quantitatively evaluate the pressure exerted by discharges from large-scale wastewater treatment plants (WWTP). For the calculation of this analysis, it will include a general approach to the state of the system, as well as a more detailed assessment of each of the surface water bodies that make up the

WIT Transactions on Ecology and the Environment, Vol 234, © 2019 WIT Press
www.witpress.com, ISSN 1743-3541 (on-line)
doi:10.2495/RBM190111

fluvial network of the case study. In addition, changes in the vulnerability of ecosystems to this index during droughts and in future Climate Change scenarios will be analysed.

2 STUDY CASE

The study was carried out within the framework of the Catalan Internal Basins, a Spanish hydrographic demarcation located in the Northwest of the Iberian Peninsula (Fig. 1). The territorial scope of CIC includes the basins of the rivers Muga, Fluvià, Ter, Daró, Tordera, Besòs, Llobregat, Foix, Gaià, Francolí and Riudecanyes, and the coastal streams between the border with France and the outlet of the river Sénia. The set of internal basins is organised into 28 hydrological units, which are subdivided into 261 surface water bodies [9].

One of the peculiarities of this system is the importance of urban demand in relation to other consultative uses of water such as agriculture or industry. In total, 55% of the extractions are destined to urban supply. This characteristic has motivated the decision to use the CIC as a pilot basin for the application of this analysis. It is assumed that urban returns from WWTP discharges will have special relevance in the total water balance of the system.

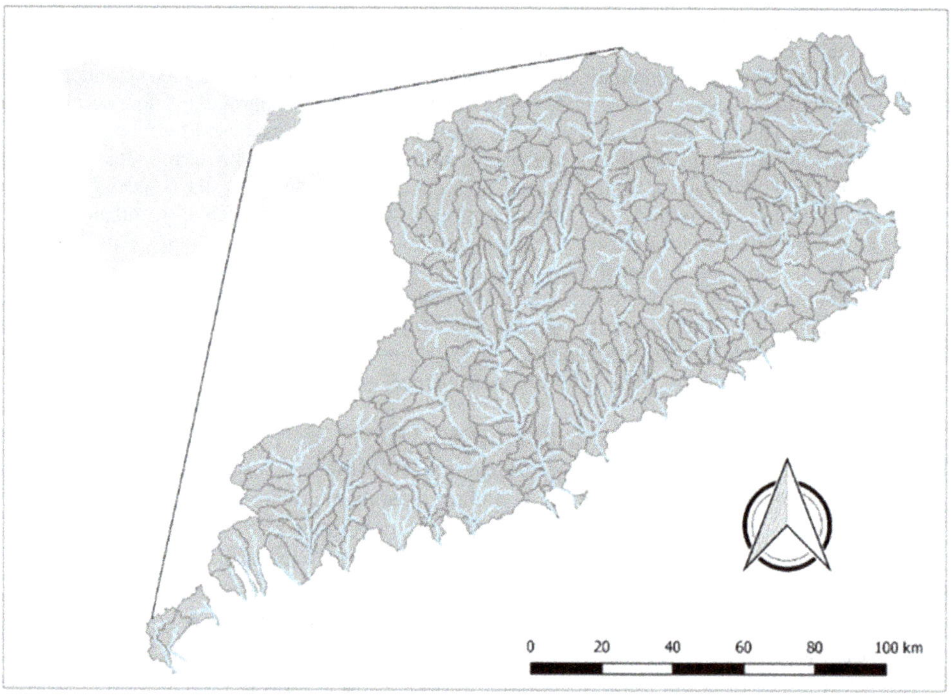

Figure 1: Location of Catalan Internal Basins and 261 sub-watersheds.

3 DATA AND METHODOLOGY

3.1 Data

Part of the data used for the development of this work has been provided by the Catalan Water Agency (ACA), the governmental body responsible for the management of the CIC. Among the information provided, they stand out:

1. Inter-basin inflows for the period Oct/1980–Sep/2008, calculated using the Sacramento Soil Moisture Accounting (SSMA) model.
2. Database of registers of authorised dumping dossiers located in the study area.
3. Location of the most relevant water extraction points of the system for urban and agricultural supply.
4. Records of outflow from reservoirs located within the demarcation.
5. Database with a series of flows observed by the ACA's network of gauges.

In addition, information was downloaded from the European project SWICCA on the possible impacts to which the study area can be subjected to different scenarios of Climate Change. Of all the parameters offered by the SWICCA platform, the monthly percentage change has been used at the level of inflows by sub-basins. Table 1 shows the regional climate models for which the results have been used for calculating the average ratio change of the three Climate Change scenarios included in this work (Table 2).

Table 1: Regional climate models by emissions scenario.

Model by emissions scenario		
RPC2.6	**RPC4.5**	**RPC8.5**
• CSC_REMO2009_MPI-ESMLR_rcp26 • SMHI_RCA4_EC-EARTH_rcp26	• KNMI_RACMO22E_EC-EARTH_rcp45 • SMHI_RCA4_HadGEM2-ES_rcp45 • CSC_REMO2009_MPI-ESM-LR_rcp45 • IPSL_WRF33_CM5A_rcp45 • SMHI_RCA4_EC-EARTH_rcp45	• KNMI_RACMO22E_EC-EARTH_rcp85 • SMHI_RCA4_HadGEM2-ES_rcp85 • SMHI_RCA4_EC-EARTH_rcp85 • CSC_REMO2009_MPI-ESM-LR_rcp85

Table 2: Percentage annual average change in inflows by zone.

	Annual Mean (% inflow change)		
Catalonia Internal Basins areas	**2020**	**2050**	**2080**
Coast Basins	5.82	4.82	7.00
Upper Llobregat	-3.73	-5.36	-8.82
Lower Llobregat	0.25	-0.95	-2.53
Upper Ter	0.45	0.09	-2.55
Lower Ter	6.53	4.43	3.32

The study case has zoned according to the basin characteristics, in order to adjust, as far as possible, the impacts of the three Climate Change scenarios on the monthly inflows (Fig. 2).

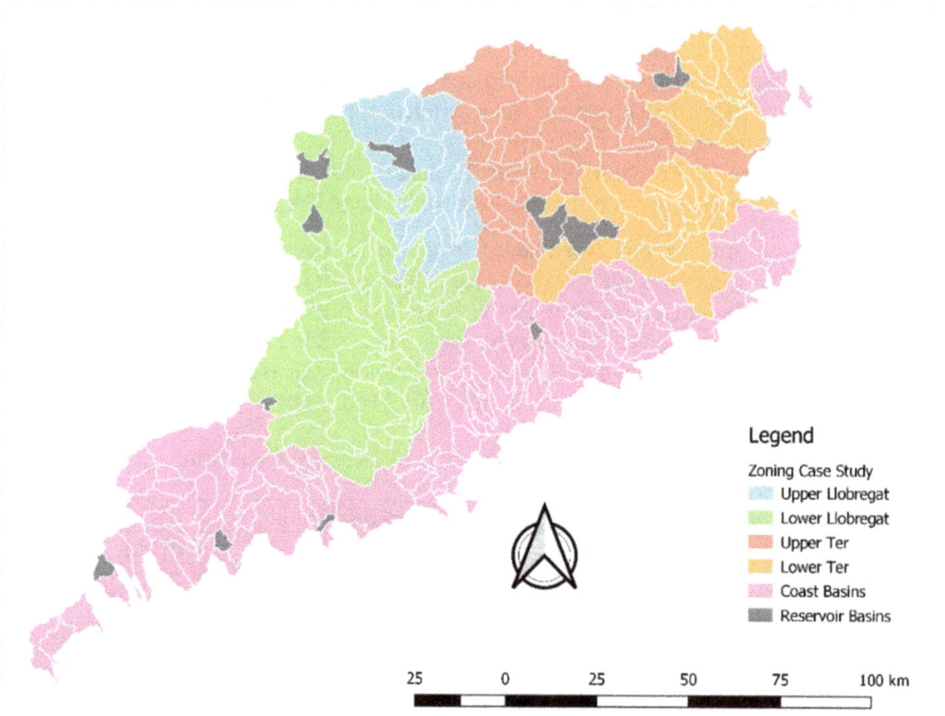

Figure 2: The study area zoning.

3.2 Methodology

In order to carry out the study, it was necessary to have a water balance of the entire system studied, which includes the main elements that directly affect the natural fluvial regime. The Rapid Response Tool of the Environmental State [R^{2}EA] was used for its development. This model was developed by the water resources research group of the Universitat Politècnica de València and is aimed at quantitatively and qualitatively evaluating the effect of a large-scale strut and diffuse pollution. The present work has focused on evaluating the quantitative results offered by the model.

The R^{2}EA tool requires the input data to be in a format and with specific units, so first a pre-treatment of data provided by the water authority was applied.

3.2.1 Inflows

The sub-basins used by the Sacramento method did not coincide with the classification of water bodies carried out by the ACA. Therefore, a first spatial extrapolation was made, through which the percentage of the Sacramento basin that coincided with the area of the ACA basins was assigned. The QGIS geospatial tool was used to carry out this step.

Then, the inflows from the Sacramento basins were distributed to the ACA basins. Finally, it was verified that the allocation had been correctly executed by checking whether the annual balance (1995–2008 period) of two head watersheds conformed to the data observed from the gauging stations downstream of them (Fig. 3).

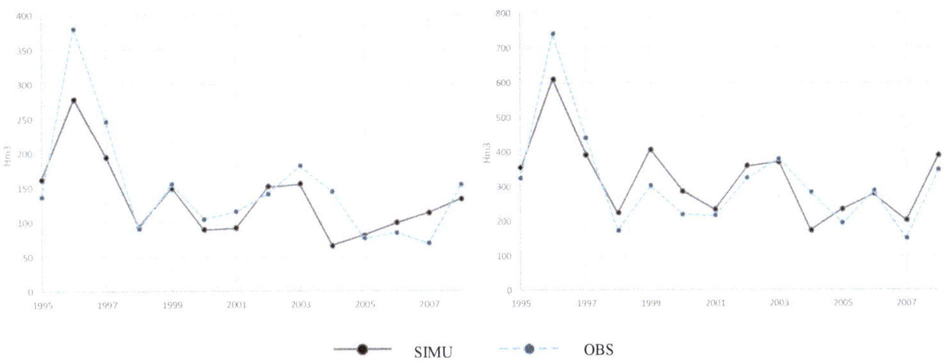

Figure 3: Check between simulated and observed flows.

3.2.2 Discharges from WWTPs

Once the inflows were adopted, the information in the discharge files was analysed. First, a geospatial analysis was performed using GIS to identify the masses affected by urban discharges. In this case, the geographic information system used was QGIS.

Next, the summation of the volume discharged per body of water was made, since in some cases the bodies of water are affected by more than one WWTP discharge. After this, a matrix was obtained where the total volume discharged by the water body and month are collected.

3.2.3 Flow regulation and extraction

In order to model the water balance in an altered regime and not in the natural one, the flows downstream of the main reservoirs were corrected using the outflow registers. In addition, the masses affected by the most important strut demands were located. In total, information was introduced on the regulation of five reservoirs (Susqueda, Darnius, Boadella, Foix, Sant Ponç and la Baells) and flow extractions were included for a total of 520.3 Hm^3.

3.2.4 Simulations:

Once the base model R^2EA was built, the two first simulations carried out were executed by Climate Change scenario (2020, 2050, 2080).

In the first one, the returns were not considered, so only the natural flows (Q_{nat}) are taken into account.

In the second, the water inputs to the WWTP discharge system (Q_{disch}) were included. Subsequently, the relationship between the total circulating flow (Q_{total}) and the circulating flow per discharge (Q_{disch}) was calculated, both in the current scenario and in the Climate Change scenarios.

$$Q_{total} = Q_{nat} + Q_{disch}, \tag{1}$$

$$\%\, Q_{disch} = \frac{Q_{total} - Q_{nat}}{Q_{total}}. \tag{2}$$

Finally, the global analysis of the state of the demarcation and the individual evaluation of each mass with respect to the calculated parameter were carried out.

4 RESULTS

4.1 Percentage of streamflows from urban discharges (% Q_{disch}).

Fig. 4 shows the average relationship between the total flow and the flow from urban discharges of all the water bodies affected. We have identified 149 masses directly affected by discharges and 24 more that are indirectly affected. Fig. 4 shows a significant increase in the relationship between the natural flow and the flow from urban discharges during the summer months (June to October). In addition, the months with the highest and lowest discharge pressure risk are September and February, respectively. Finally, there are no significant differences between the global results obtained for the three Climate Change scenarios.

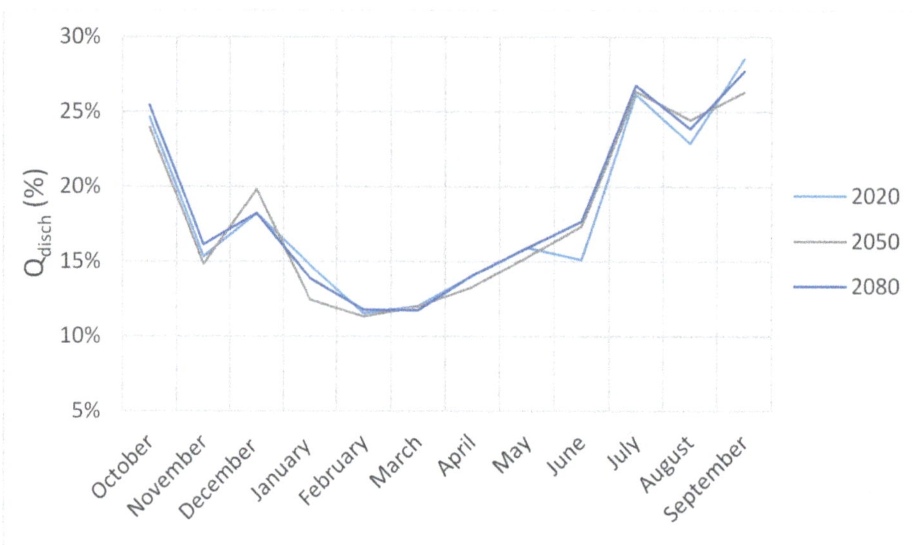

Figure 4: Annual variation of the index (Q_{disch}) for the CIC.

Then, the annual variation was analysed by zones, and is illustrated in Fig. 5: Upper Llobregat (Fig. 5(a)); Lower Llobregat (Fig. 5(b)); Upper Ter (Fig. 5(c)); Lower Ter (Fig. 5(d)) and Coast Basins (Fig. 5(e)). Firstly, the results show an index increase between upper and lower basins of the two main rivers, Llobregat and Ter. This difference reaches values of as much as 15 points higher in the low areas than in the high areas during the same month. This situation is because it is in the lower part of the rivers, near the coast, where most of the population living in the Catalan Internal Basins is located.

The most significant case is that of the coastal basins, characterized by irregular regimes, which are aquatic ecosystems very vulnerable to both quantitative and qualitative alterations. As the results show, in this case, the average streamflows from urban discharges reach 40% of the total during the summer months.

Figure 5: Annual variation of the index (Q_{disch}) by zone. (a) Upper Llobregat; (b) Lower Llobregat; (c) Upper Ter; (d) Lower Ter; (e) Coast Basins.

Finally, no significant impacts have been detected among the Climate Change scenarios presented in any of the areas studied. Therefore, the results in this sense are not conclusive.

In order to identify the masses that suffer the greatest pressure, the maximum values registered during the simulated period were presented per mass of water, in map format (Fig. 6). The maximum value is represented because it is related to the most critical situation that each body of water can reach.

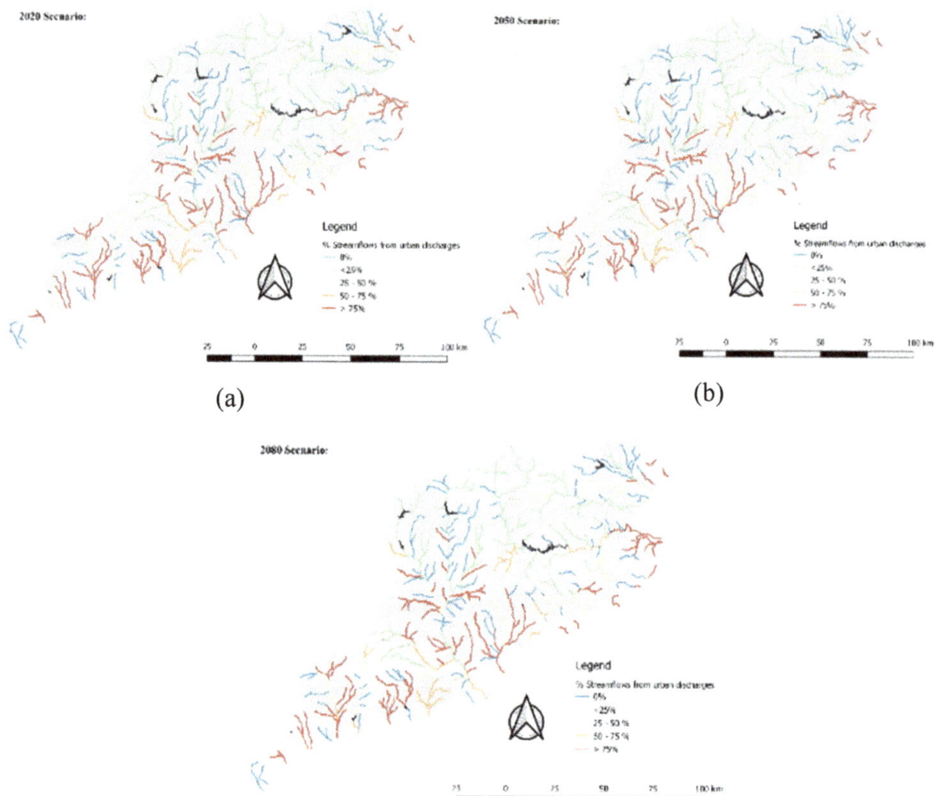

(a) (b)

(c)

Figure 6: Maximum value of the index (Q_{disch}) during the simulated period by water body. (a) 2020 Scenario; (b) 2050 Scenario; (c) 2080 Scenario.

As in Fig. 5, a higher risk is identified in the lowlands and coastal areas where the pressure from urban discharges is higher. However, when presenting the results in map format, it is observed that most of the rivers for which the maximum value of circulating flow from WWTPs exceeds 75% are secondary rivers. These rivers are characterized by low and intermittent flows.

On the other hand, when the maps are compared with each other, there are no appreciable changes in the level of risk between the three Climate Change scenarios. Table 3 shows the number of masses classified in each group by scenario. There has been a minor improvement in the quantitative risk of urban discharges between 2020 and 2050 because, in the second scenario, a small increase in total inflows is expected. Finally, in the 2080 scenario, there was an increase in the number of masses that were in the range of 50–75%.

Table 3: Number of water bodies by sub-class of Q_{disch} and Climate Change scenario.

No. of water bodies by % group	2020	2050	2080
>75%	68	67	65
75–50%	11	9	16
50–25%	32	35	31
<25%	62	62	61

4.2 The percentage of demands from WWTP discharges

The model includes 14 water extraction points for agricultural and urban demands, 520 Hm^3 per year. Fig. 7 presents the average percentage of the flow that comes from discharges upstream of each abstraction. It can be observed how the water bodies ES100MSPF2000420 and the ES100MSPF2000460 reach average values higher than 25%.

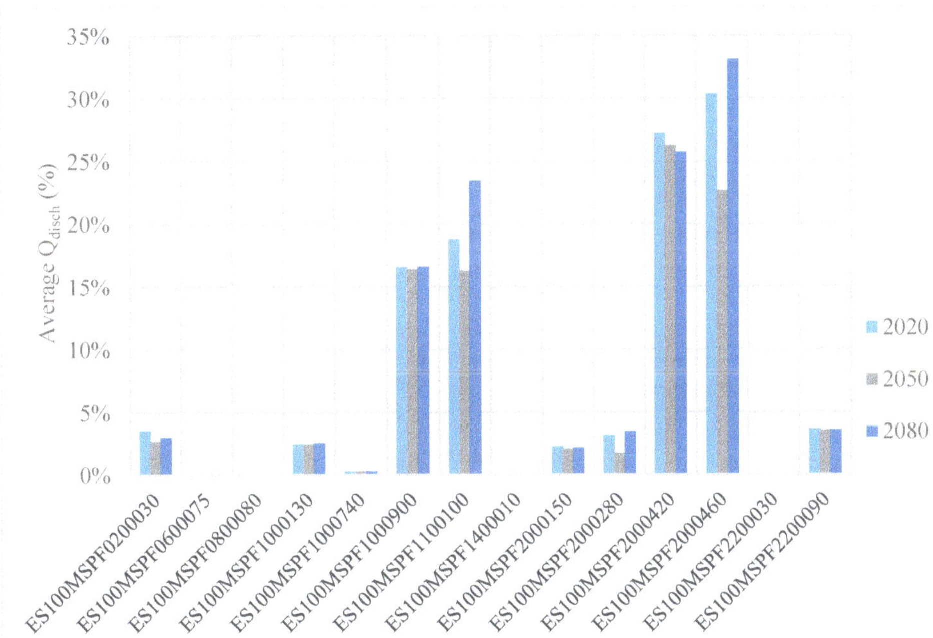

Figure 7: Average Q_{disch} of water bodies affected by important demands.

Finally, no clear overall trend has been identified as to how Climate Change may affect the case study.

5 CONCLUSION

Although the main objective of the study was to identify the effect of Climate Change on the relationship between the volumes discharged by WWTPs and natural flows, it did not identify a general trend. This is due to the fact that the SWICCA information about changes in rates of inflows also does not follow a clear pattern. The streamflows increase or decrease depending on the analysed area, so the total water balances are compensated.

A general trend has been observed by which the vulnerability of the analysed masses increases during the summer months. The reduction of flows and water stress situations suppose a challenge for basin managers, since the risk to which these aquatic ecosystems are subjected increases considerably. This trend increases when comparing the evolution of the indicator in each zone. Coastal areas characterized by higher population densities are subjected to greater pressure from urban discharges.

In addition, in the present study, the important role of the water that comes from urban discharges in the hydraulic balance has been demonstrated. The high anthropization level of the rivers, due necessary to take into account the role of urban discharges on the water river systems.

In summary, the R^2EA model allows the opportunity for the water authorities to know the relation between urban discharges flows and total streamflows as an ecological indicator, to characterize river basins pressures, to analyse annual vulnerability variation and to simulate large periods, in order to identify the most critical situation of each of the water bodies.

ACKNOWLEDGEMENTS

The authors thank the Spanish Research Agency (MINECO) for their financial support to the ERAS project (CTM2016-77804-P, including EU-FEDER funds).

REFERENCES

[1] Pellicer-Martínez, F. & Martínez-Paz, J.M., The Water Footprint as an indicator of environmental sustainability in water use at the river basin level. *Sci. Total Environ.*, **571**, pp. 561–574, 2016.

[2] Martínez-Paz, J., Pellicer-Martínez, F. & Colino, J., A probabilistic approach for the socioeconomic assessment of urban river rehabilitation projects. *Land Use Policy*, **36**, pp. 468–477, 2014.

[3] Pellicer-Martínez, F. & Martínez-Paz, J.M., Grey water footprint assessment at the river basin level: Accounting method and case study in the Segura River Basin, Spain. *Ecol. Indic.*, **60**, pp. 1173–1183, 2016.

[4] Pophare, A.M., Lamsoge, B.R., Katpatal, Y.B. & Nawale, V.P., Impact of over-exploitation on groundwater quality: A case study from WR-2 watershed, India. *J. Earth Syst. Sci.*, **123**(7), pp. 1541–1566, 2014.

[5] Foster S. et al., Quaternary aquifer of the North China Plain: Assessing and achieving groundwater resource sustainability. *Hydrogeol. J.*, **12**(1), pp. 81–93, 2004.

[6] Sultana, M.N., Uddin, M.M., Ridoutt, B.G. & Peters, K.J., Comparison of water use in global milk production for different typical farms. *Agric. Syst.*, **129**, pp. 9–21, 2014.

[7] Rijsberman, F.R., Water scarcity: Fact or fiction?. *Agric. Water Manag.*, **80**(1–3), Special Issue, pp. 5–22, 2006.

[8] Sandoval-Solis, S., McKinney, D.C. & Loucks, D.P., Sustainability Index for Water Resources Planning and Management. *J. Water Resour. Plan. Manag.*, **137**(5), pp. 381–390, 2010.

[9] Generalitat de Catalunya & Agència Catalana de l'Aigua, Pla de gestió del districte de conca fluvial de Catalunya (2016–2021). Government agreement issued by Department of Territory and Sustainability, 2017.

FINANCING OF WATER POLICY: COST RECOVERY OF THE HYDROGRAPHIC CONFEDERATIONS OF SPAIN

MARCOS GARCÍA LÓPEZ, BORJA MONTAÑO SANZ & JOAQUÍN MELGAREJO MORENO
University Institute of Water and Environmental Sciences, University of Alicante, Spain

ABSTRACT

The shortage of water resources and the complex institutional framework in Spain have originated several different organizations responsible for devising water policy measures that have to share the funding. With this distribution of functions, each organization is responsible for some tasks and with funding to perform them. However, taking into account that in Spain the full financing of all aspects of water policy is not achieved, it is understandable that the Hydrographic Confederations are not capable of recovering the costs of their activities. We must add that the distribution of financial resources may not be the most appropriate, which would leave the Hydrographic Confederations in a complicated situation, this paper intends to examine the current financial situation of these organizations. The results of the analysis show that there is a lack of recovery of financial costs in the main activities, which would be offset by a transfer of external funds to the water policy, since this must necessarily be financed. This also implies that, if the environmental costs are added to the financial costs, the result is a very low cost recovery. In addition, there are activity plans that are not being carried out due to the lack of financing, so it seems appropriate to suggest an increase in prices or water pricing to compensate for this lack of funding.

Keywords: Hydrographic Confederations, cost recovery, economic cost, environmental cost, financing of water policy.

1 INTRODUCTION

Recently, water has ceased to be considered a renewable and infinite resource because, although it is from the natural cycle of water, this feature of the resource has a limit and can become a non-renewable resource if its exploitation is done irresponsibly. In addition, its geographical and temporary distribution is a major problem in Spain, since there are significant imbalances between available resources and water needs in different geographical areas [1].

The increase registered by water supply provisions, together with the increase in population experienced by many urban areas, have made traditional resource sources insufficient [2]. In this way, the development of new sources is key to ensure the sustainability of a fundamental resource such as water, as they increase their availability while reducing the environmental impact derived from the management of water sources.

Currently there are high water stress problems in many parts of the world, including Spain, where there is a considerable dispersion in the distribution of water resources, which leads to great scarcity in a relevant part of the territory and abundance in a few zones. When demand exceeds available resources and their rate of regeneration, strong competition for resources can be expected, leading to scarcity and high pressure on resources, thus increasing the likelihood of conflicts and management and pollution problems, having a negative impact on health and economic activity. Therefore, the growing scarcity leads to greater vulnerabilities of the economy in situations of drought and to a growing deterioration in the environmental quality of surface and underground water bodies [1].

Another important aspect to assess is that the incidence of droughts in Europe is not stable, but has increased by 20% since 1976 and, despite the increase in torrential rains, annual rainfall has to fall [3]. The overexploitation of the aquifers must be added to the droughts and

WIT Transactions on Ecology and the Environment, Vol 234, © 2019 WIT Press
www.witpress.com, ISSN 1743-3541 (on-line)
doi:10.2495/RBM190121

precipitations, so that the need to take water policy measures is evident, since water is a fundamental resource and its correct exploitation is at the base of sustainable development, besides being social demand base [4].

These problems make the situation difficult to solve and even worse when taking into account that in Spain there is a large institutional network. The different organisms that are part of this network have different areas of activity, in particular, we have international, European, national, regional, provincial and local institutions, in addition to the hydrographic confederations and others such as the irrigation communities.

The presence of so many institutions leads to problems of overlapping of competences and functions, conflicts, dilution of responsibilities and shows tendency to create new ones instead of modifying existing ones [5]. These problems, derived from the large number of bodies with responsibilities in terms of water policy, lead to inefficiency in management of water resources. In addition, in line with the objective of this work, the existence of so many entities, with their distribution of competences, implies the need to distribute the financial funds so that everyone can carry out their tasks. Therefore, the funds must be sufficient to finance all water policy measures, but they must be distributed appropriately.

The hydrographic confederations are a fundamental part of the aforementioned institutions, which have diverse and very important functions, for which reason their adequate financing is fundamental for the correct elaboration and execution of water policy. Therefore, this paper will analyse the economic situation of these organizations after commenting on the importance of cost recovery.

2 IMPORTANCE OF COST RECOVERY

Water policy is of fundamental importance in society, since water is a basic resource without which could not continue running. This makes water policy a key aspect as it is responsible for the management of water resources and, like any other policy, has an associated high cost that must be financed. When analysing these costs, a distinction must be made between two types, some derived from the need for financial resources to be able to carry out the different activities, while others arise from the need to combat the environmental impact of water policy executed, that is, you can differentiate between economic and environmental costs. As is logical, the new policies must be cautiously elaborated to carry them out in the most efficient way possible, so that their economic cost, but also environmental cost, can be minimized.

The economic costs do not require much explanation, because like any activity, the water policy has an economic cost and requires financing to be able to run. However, the environmental costs are a more complex problem, which is why the latter are going to be treated.

The impact on the natural environment is defined by the Water Framework Directive (WFD), which establishes that an activity, to the extent that it influences on the state of the ecosystems, has an environmental impact that must be corrected. This directive, in the application of "polluter pays", establishes the principle of cost recovery, both economic and environmental [6]. With this, the European Union emphasise the importance of adequate financing of water policy, so that the water supply is economically and environmentally sustainable.

One of the main problems in terms of environmental costs is the great difficulty in calculating them, since it is very complicated to measure the value of natural resources due to their economic, environmental, social and cultural character [5]. In addition, the forms that the environmental impact can adopt are very varied and can affect differently to each type of value, which complicates the development of appropriate measures for their correction. As an example, Yongguan et al. [7] analysed the environmental cost of water pollution in the Chinese city of Chongqing, finding that it has significant effects on health, plants, fish, animals and industry, from which is derived not only a negative environmental effect, but also economic losses.

However, there is a relevant aspect that should not be forgotten about the environmental objectives of water policy. The WFD, in search of economic and use of resources efficiency, gives the possibility of eliminating this type of objectives if the economic cost necessary to comply with it is too high for the benefit of achieving it, thus placing more emphasis on the efficient performance of the analyses and actions that are required. In addition, the measures considered must be well received by the public, that is, public preferences must be included in the analysis and decision-making processes [8].

Even if the environmental benefits are taken into account, the economic costs must be financed and the environmental impact must be counteracted, so the economic income is necessary for the water policy, which comes from the price of water and the taxes on water consumption, which is reflected in the water bill. In addition, these revenues must be properly distributed among the different agencies that have assigned water policy responsibilities.

The DMA establishes the recovery of all costs as mandatory, since it is considered as a key to the sustainability of water resources. In terms of economic costs, the financing of these is necessary so that the water policy can continue to be implemented, while the environmental ones suppose a damage on the natural environment that we can't ignore and allow it to accumulate, but repair it to enjoy an environment in good conditions.

However, talking about the economic costs, the WFD does not seek to establish a ranking of cost recovery and a strict criterion, but the objective is to provide the water policy with the necessary resources to carry it out [9]. One aspect to highlight is about who is contributing, since must be the users who enjoy the water services who must pay for its maintenance, while those responsible for the environmental damage must bear the cost of its correction. However, if full costs recovery is not achieved through the usual income from water services, the remaining funding will come from users who are paying in another way, so a tax adjustment would be advisable to make financing fairer.

In order to study the current situation of cost recovery, in the next two sections will be used the cost recovery chapters or annexes of the Spanish inter-community hydrographic confederations for the hydrological cycle 2015–2021 and the recently published documents, in public consultation phase, of the plans for the 2021–2027 cycle, thus being able to compare between both plans.

3 COST RECOVERY IN THE SPANISH INTER-COMMUNITY HYDROGRAPHIC CONFEDERATIONS

In this section are analysed the cost recovery data of the inter-community Hydrographic Confederations of Spain for the hydrological cycle 2015–2021, in order to examine the financial situation in this period. Since all the inter-community confederations of the country are included, the recovery data are very high and only those related to the totality of the water services will be included in this study, without specifying the level of recovery in each of them. In addition, to consider the effect of environmental impact, cost recovery is shown in terms of recovery of financial and total costs.

The available data are shown in Table 1, reflecting that an important part of the financial costs is pending to recover and that the level of total costs recovery is far from 100% when taking into account the environmental and the resource ones. Specifically, it can be seen that the highest level of cost recovery comes from the Tajo Hydrographic Confederation with 91% of the financial and 81% of the total. For its part, the Miño-Sil Hydrographic

Confederation only recovers 38% of the financial and 34% of the total costs, which not only represents an important difference compared to the majority, but in this case the cost recovery is really low.

Table 1: Recovery of costs, total and financial, of the Spanish Inter-community Hydrographic Confederations for the hydrological cycle 2015–2021. *(Source: Own elaboration based on data from the Inter-community Hydrographic Confederations.)*

Plan 2015–2021		
Hydrographic Confederation	Financial costs	Total costs
Cantábrico Occidental	85%	76%
Cantábrico Oriental	74%	67%
Ceuta	69%	56%
Melilla	41%	40%
Duero	64%	49%
Ebro	76%	65%
Guadalquivir	83%	74%
Guadiana	80%	59%
Tajo	91%	81%
Júcar	84%	78%
Segura	83%	57%
Miño-Sil	38%	34%

Note: Information regarding the data used in this table can be found in the Appendix.

The rest of the confederations are distributed among the levels of cost recovery of the Tajo and the Miño-Sil, but, in general, closer to the level of the former. Taking into account these confederations, the unweighted average recovery of financial and total costs is 72% and 61%, respectively. Therefore, there is a reduction in the level of cost recovery of 11 points by adding environmental costs to financial ones, which shows their relevance. However, this change is, again, an unweighted average, so there are confederations with larger and lower variations. Among these variations we can highlight the percentage point of Melilla and the 4 of the Miño-Sil which, although they are the two confederations with the least cost recovery, are the ones that see their situation less worsened by incorporating environmental costs. On the other hand, the large variations of 21 and 25 percentage points of Guadiana and Segura, respectively, are important.

The results, obtained by the confederations themselves, show a lack of capacity to finance the water policy carried out in Spain, even some of these organizations have prepared additional water policy measures that are not being implemented due to lack of income. This fact leads, as mentioned above, to the fact that the current situation of water resources cannot be considered sustainable, since the water policy is not being fully financed and does not adequately address the environmental impact of human water resources management. In addition, it must be kept in mind that the water policy must necessarily be financed, so that in order to complete the cost recovery, funds will be received from other ways independent of water services. In this way, people other than consumers are paying to maintain the situation, which implies that the polluter pays principle is not being met.

It must also be considered that this lack of funding may be blocking the execution of additional measures and the internalization of environmental costs. As explained by the Segura Hydrographic Confederation [10]: "there is a pending environmental cost (future) and another already internalized, corresponding to the measures already executed and included in the financial costs". In other words, the environmental costs are not fully internalized because certain measures cannot be implemented, hence the distinction between financial and environmental costs of the data.

4 COMPARISON OF THE CURRENT PLANS WITH THE PUBLISHED DATA OF THE NEXT ONES

In this section, taking advantage of the recent publication last October of the initial documents of hydrological planning for the 2021–2027 period, a comparison is made between the levels of cost recovery of the current cycle with those of the next. Table 2 shows the recovery levels of both financial and total costs for both hydrological cycles, showing first the financial ones and, second, the totals of both cycles.

Table 2: Cost recovery, total and financial, of the Spanish inter-community hydrographic confederations for the hydrological cycles 2015–2021 and 2021–2027. *(Source: Own elaboration based on data from the Inter-community Hydrographic Confederations.)*

Cost recovery*	Financial costs		Total costs	
Hydrographic Confederations	2015–2021	2021–2027**	2015–2021	2021–2027**
Cantábrico Occidental	85%	86%	76%	77%
Cantábrico Oriental	74%	79%	67%	74%
Ceuta	69%	68%	56%	66%
Melilla	41%	37%	40%	37%
Duero	64%	82%	49%	49%
Ebro	76%	83%	65%	70%
Guadalquivir	83%	84%	74%	79%
Guadiana	80%	71%	59%	50%
Tajo	91%	88%	81%	73%
Júcar	84%	93%	78%	87%
Segura	83%	82%	57%	63%
Miño-Sil	38%	49%	34%	47%

Notes: * Data is rounded. ** In some cases, in the documents under public consultation for the planning of the next hydrological cycle (2021–2027) data on cost recovery of the current cycle have varied slightly, but it has been chosen to maintain the memory data of the current hydrological plan because in these new documents only the data of total costs recovery of the current plans are offered, leaving out the distinction between financial and environmental costs. Information regarding the data used in this table can be found in the Appendix.

In general, the conclusions obtained through the data of the current period can be extended to the new data, although with some differences. Taking into account the change in recovery of financial costs from one period to another, we find both positive and negative variations, with 7 confederations that improve their situation compared to 5 that worsen, with 3 percentage points as unweighted average of these variations. Thus, we can see important improvements like the 18 points of the Duero and the 11 points of the Miño-Sil, but also setbacks like the 9 points of the Guadiana, which is the only place with such a high negative change.

On the other hand, taking the total costs the result is similar because now 8 confederations improve their situation by 3 that make it worse, while one maintains the total costs recovery in both cycles, being now the average of variation the same as for financial costs. In this case, we find again significant improvements, highlighting the 13 points of the Miño-Sil and the 10 points of Ceuta, but there are two strong negative variations, specifically the Guadiana and the Tajo with variations of 9 and 8 points, respectively. These last two, together with Melilla, which drops 3 points, are the only ones that see worsened their situation.

However, one aspect to note is that there are three confederations with opposite variations in cost recovery. In the first place, Ceuta worsened its financial costs recovery at one point, but improved the recovery of the totals by 10 points, showing a better environmental situation. Second, the Duero, with an improvement of 18 points in the financial costs recovery, has a variation of tenths in terms of totals. This is due to a greater decrease in financial costs than in revenues, however, the recovery of total costs barely varies due to the reduced improvement in the environmental cost, which acquires a greater weight in the accounts. Finally, the Segura, with a 1 point reduction in financial costs recovery, obtains an improvement of 7 points in terms of the total costs recovery.

Again, the data of the confederations themselves show inability to recover financial and environmental costs, in addition to the internalization of the latter, which compromises the sustainability of water resources and allows, again, to propose an increase in the price of water. water services or another source of income for water policy.

5 CONCLUSIONS

Water resources are currently a scarce resource that requires a large number of water policy measures to manage it efficiently. This is of great importance, because water is a fundamental resource without which human activity could not develop. Of course, these measures have an economic cost that must be financed by necessity, but also the environmental impact that human activity has on the environment must be addressed, so an additional economic cost must be considered depending on the one that has to repair the damage caused. The recovery of these costs is necessary to be able to consider the water policy as sustainable.

To execute the measures, in Spain there is a large number of institutions that share the functions and income, so that each of them can carry out its activity. Therefore, the proper distribution of financing between institutions is as important as obtaining it. Observing the data of the Spanish inter-community hydrographic confederations, it can be concluded that cost recovery is not being achieved, without even reaching the recovery of financial costs, so that the activity currently is not sustainable and there are still important efforts to be made. To conclude more accurately that the recovery of water policy costs is not being achieved it would be necessary to analyse a greater number of institutions, but it can be concluded that no Spanish inter-community hydrographic confederation recovers them, reflecting a structural problem in this area.

In sum, water policy at the river basin level is not adequately financed and requires more funds from water resources to avoid obtaining them from other places. The latter is relevant because service users are those who must pay for it instead of other individuals or agents. As a recommendation, it seems appropriate to propose an increase in the price of water and/or in the taxes levied on its consumption to improve the situation, since the current price of water does not respond to the scarcity and importance of this resource.

APPENDIX

- Confederación Hidrográfica de Ceuta, Memoria. Plan hidrológico de la Demarcación Hidrográfica de Ceuta, 2015. www.chguadalquivir.es/demarcacion-hidrografica-ceuta#Segundociclodeplanificaci%C3%B3n(2016-2021). Accessed on: 7 Feb. 2019.

- Confederación Hidrográfica de Ceuta, Documentos iniciales. Memoria. Plan hidrológico de la Demarcación Hidrográfica de Ceuta. Revisión de tercer ciclo (2021–2027), 2018. www.chguadalquivir.es/demarcacion-hidrografica-ceuta. Accessed on: 7 Feb. 2019.
- Confederación Hidrográfica de Melilla, Memoria. Plan hidrológico de la Demarcación Hidrográfica de Melilla, 2015. www.chguadalquivir.es/demarcacion-hidrografica-melilla#Segundociclodeplanificaci%C3%B3n(2016-2021). Accessed on: 7 Feb. 2019.
- Confederación Hidrográfica de Melilla, Documentos iniciales. Memoria. Plan hidrológico de la Demarcación Hidrográfica de Melilla. Revisión de tercer ciclo (2021–2027), 2018. www.chguadalquivir.es/demarcacion-hidrografica-melilla. Accessed on: 7 Feb. 2019.
- Confederación Hidrográfica del Cantábrico, Anejo 09. Recuperación de costes de los servicios del agua. Plan hidrológico de la Demarcación Hidrográfica del Cantábrico Occidental 2015–2021, 2015. www.chcantabrico.es/dhc-occidental. Accessed on: 7 Feb. 2019.
- Confederación Hidrográfica del Cantábrico, Memoria. Plan hidrológico de la parte española de la Demarcación Hidrográfica del Cantábrico Oriental, 2015. www.chcantabrico.es/parte-espaniola-de-la-dhc-oriental. Accessed on: 7 Feb. 2019.
- Confederación Hidrográfica del Cantábrico, Documentos iniciales. Memoria. Plan hidrológico de la Demarcación Hidrográfica del Cantábrico Occidental. Revisión de tercer ciclo (2021–2027), 2018. www.chcantabrico.es/planes-hidrologicos-2021-2027/dhc-occidental/documentos-iniciales. Accessed on: 7 Feb. 2019.
- Confederación Hidrográfica del Cantábrico, Documentos iniciales. Memoria. Plan hidrológico de la parte española de la Demarcación Hidrográfica del Cantábrico Oriental. Revisión de tercer ciclo (2021–2027), 2018. www.chcantabrico.es/planes-hidrologicos-2021-2027/dhc-oriental/documentos-iniciales. Accessed on: 7 Feb. 2019.
- Confederación Hidrográfica del Duero, Memoria de Plan hidrológico 2015–2021. Plan hidrológico de la parte española de la Demarcación Hidrográfica del Duero, 2015. www.chduero.es/Inicio/Planificaci%C3%B3n/Planhidrol%C3%B3gico20152021/Plan Hidrol%C3%B3gico/tabid/734/Default.aspx. Accessed on: 7 Feb. 2019.
- Confederación Hidrográfica del Duero, Documentos iniciales. Memoria. Plan hidrológico de la parte española de la Demarcación Hidrográfica del Duero. Revisión de tercer ciclo (2021–2027), 2018. www.chduero.es/Inicio/Planificaci%C3%B3n/Planhidrol%C3%B3gico20222027Propuesta/Documentosiniciales/tabid/767/Default.aspx. Accessed on: 7 Feb. 2019.
- Confederación Hidrográfica del Ebro, Memoria. Plan hidrológico de la parte española de la Demarcación Hidrográfica del Ebro, 2015. www.chebro.es/contenido.visualizar.do?idContenido=42695&idMenu=4780. Accessed on: 7 Feb. 2019.
- Confederación Hidrográfica del Ebro, Documentos iniciales. Memoria. Plan hidrológico de la parte española de la Demarcación Hidrográfica del Ebro. Revisión de tercer ciclo (2021–2027), 2018. www.chebro.es/contenido.visualizar.do?idContenido=56837&idMenu=5781. Accessed on: 7 Feb. 2019.
- Confederación Hidrográfica del Guadalquivir, Memoria. Plan hidrológico de la Demarcación Hidrográfica del Guadalquivir, 2015. www.chguadalquivir.es/demarcacion-hidrografica-guadalquivir#Segundociclodeplanificaci%C3%B3n(2016-2021). Accessed on: 7 Feb. 2019.
- Confederación Hidrográfica del Guadalquivir, Documentos iniciales. Memoria. Plan hidrológico de la Demarcación Hidrográfica del Guadalquivir. Revisión de tercer ciclo

(2021–2027), 2018. www.chguadalquivir.es/demarcacion-hidrografica-guadalquivir. Accessed on: 7 Feb. 2019.
- Confederación Hidrográfica del Guadiana, Memoria (Parte II). Plan hidrológico de la Demarcación Hidrográfica del Guadiana, 2015. http://planhidrologico2015.chguadiana.es/?corp=planhidrologico2015&url=61. Accessed on: 7 Feb. 2019.
- Confederación Hidrográfica del Guadiana, Documentos iniciales. Memoria. Plan hidrológico de la Demarcación Hidrográfica del Guadiana. Revisión de tercer ciclo (2021–2027), 2018. www.chguadiana.es/planificacion/plan-hidrologico-de-la-demarcacion/revision-del-plan-hidrologico-2021-2027/documentos-iniciales. Accessed on: 7 Feb. 2019.
- Confederación Hidrográfica del Júcar, Memoria. Plan hidrológico de la Demarcación Hidrográfica del Júcar, 2015. www.chj.es/es-es/medioambiente/planificacionhidrologica/Paginas/PHC-2015-2021-Plan-Hidrologico-cuenca.aspx. Accessed on: 7 Feb. 2019.
- Confederación Hidrográfica del Júcar, Documentos iniciales. Memoria. Plan hidrológico de la Demarcación Hidrográfica del Júcar. Revisión de tercer ciclo (2021–2027), 2018. www.chj.es/es-es/medioambiente/planificacionhidrologica/Paginas/PHC-2021-2027-Indice.aspx. Accessed on: 7 Feb. 2019.
- Confederación Hidrográfica del Miño-Sil, Capítulo 9. Recuperación de costes de los servicios del agua. Plan hidrológico de la Demarcación Hidrográfica del Miño-Sil, 2015. www.chminosil.es/es/chms/planificacionhidrologica/plan-hidrologico-2015-2021-vigente-rd-1-2016/80-chms/1359-plan-hidrologico-2015-2021-rd-1-2016. Accessed on: 7 Feb. 2019.
- Confederación Hidrográfica del Miño-Sil, Documentos iniciales. Memoria. Plan hidrológico de la Demarcación Hidrográfica del Miño-Sil. Revisión de tercer ciclo (2021–2027), 2018. www.chminosil.es/es/chms/planificacionhidrologica/consulta-publica-documentos-iniciales-2021-2027. Accessed on: 7 Feb. 2019.
- Confederación Hidrográfica del Segura, Documentos iniciales. Memoria. Plan hidrológico de la Demarcación Hidrográfica del Segura. Revisión de tercer ciclo (2021–2027), 2018. www.chsegura.es/chs/planificacionydma/planificacion21-27/proceso.html. Accessed on: 7 Feb. 2019.
- Confederación Hidrográfica del Tajo, Anejo 09. Recuperación de costes de los servicios del agua en la Demarcación Hidrográfica del Tajo. Plan hidrológico de la parte española de la Demarcación Hidrográfica del Tajo 2015–2021, 2015. www.chtajo.es/LaCuenca/Planes/PlanHidrologico/Planif_2015-2021/Paginas/Plan_2015-2021.aspx. Accessed on: 7 Feb. 2019.
- Confederación Hidrográfica del Tajo, Documentos iniciales. Memoria. Plan hidrológico de la parte española de la Demarcación Hidrográfica del Tajo. Revisión de tercer ciclo (2021–2027), 2018. www.chtajo.es/LaCuenca/Planes/PlanHidrologico/Planif_2021-2027/Paginas/default.aspx. Accessed on: 7 Feb. 2019.

REFERENCES

[1] Prats, D., La reutilización de aguas depuradas regeneradas a escala mundial: análisis y prospectivas. *Agua y Territorio*, **8**, pp. 10–21, 2016.

[2] Melgarejo, J., Efectos ambientales y económicos de la reutilización del agua en España. *CLM Economía*, **15**, pp. 245–270, 2009.

[3] Navarro, T., Water reuse and desalination in Spain: Challenges and opportunities. *Journal of Water Reuse and Desalination*, **8**(2), pp. 153–168, 2018. DOI: 10.2166/wrd.2018.043.

[4] Trapote, A., Tecnologías de depuración y reutilización: nuevos enfoques. *Agua y Territorio*, **8**, pp. 48–60, 2016.

[5] Torregrosa, T., *Valoración económica de los recursos naturales*. Masters thesis in Applied Economics, Universidad de Alicante, Spain, 2017.

[6] European Union, Directive 2000/60/EC of the European Parliament and of the Council of 23 October 2000 establishing a framework for Community action in the field of water policy. https://eur-lex.europa.eu/eli/dir/2000/60/oj?locale=en. Accessed on: 7 Feb. 2019.

[7] Yongguan, C., Seip, H.M. & Vennemo, H., The environmental cost of water pollution in Chongqing, China. *Environment and Development Economics*, **6**(3), pp. 313–333, 2001. DOI: 10.1017/S1355770X01000183.

[8] Alcon, F., Martin-Ortega, J., Berbel, J. & De Miguel, M.D., Environmental benefits of reclaimed water: an economic assessment in the context of the Water Framework Directive. *Water Policy*, **14**(1), pp. 148–159, 2012. DOI: 10.2166/wp.2011.001.

[9] Villar, A., Reutilización de aguas regeneradas: aproximación a los costes de producción y valoración de su uso. *Agua y Territorio*, **8**, pp. 70–79, 2016.

[10] Confederación Hidrográfica del Segura, Memoria. Plan hidrológico de la Demarcación Hidrográfica del Segura 2015–2021, 2015. www.chsegura.es/chs/planificacionydma/planificacion15-21/ Accessed on: 7 Feb. 2019.

UNDERGROUND WATERS: CHANGES IN GROUNDWATER POLICIES

FRANCINE CANSI[1] & JOAQUIN MELGAREJO MORENO[1,2]
[1]Juridical Science, Instituto Universitario del Agua y Ciencias Ambientales (IUACA), Spain
[2]History, University of Murcia, Spain

ABSTRACT

The 1988 Constitution of Brazil used the expression "water resources", not translating its economic use, but as being for the general enjoyment of the collectivity, the populace. Debates on issues related to underground water availability as another source of the water supply have raised major concerns about the quantity, quality and depletion of this resource. Even using this tuning fork, this hydrogeological body of water remains in a situation of scarcity, in the same way as the superficial waters, because the knowledge of the subject is still hindered. In Brazil, for example, the distribution of water is far from homogeneous, although there is a government effort to protect this reservoir. The greatest challenge in terms of water resources conservation lies in the lack of effective mechanisms to control, protect and use these natural resources without impeding the economic growth of the country. On the other hand, the concept of there being a water crisis goes beyond scarcity, resulting from a combination of governance failures, environmental preservation, sustainability issues and disparity in access to water. While the human right to water requires the use of this sensitive and exhaustible resource, there is an obligation for states to ensure there is secure access and secure storage, and there is a need for a more active role of sustainability to be present in society at large; however, the underground water use and management reiterate the same flaws that lead to surface water degradation. In addition, the deliberate use of groundwater requires rational actions, as resource allocation is as important in equity, which is often crucial in governance decisions. In order to understand groundwater policies, it is imperative to understand the structure and actual sequence of events by analyzing the factors or attributes that make groundwater management successful in some regions, but not others.
Keywords: global water, governance, groundwater policy, human rights, water policy, water use.

1 INTRODUCTION

The adoption of a real model for sustainable development, in particular in the management of groundwater, requires a profound change of consciousness, of the way humans perceive and understand the dependence of their life upon the natural environment; as well as lead to that sustainable model, suppressing old structures and paradigms, so that instead there is a reconstruction of a novel civilization, based on the central value of sustainability for reconstruction or transformation, in which it presents emerging options in different latitudes of the globe for the protection of waters. Baumann [1] cites for this argument that "to seek water, to wash clothes, to exchange information and opinions..." by which one agrees, because in this concept of the new demands of globalization, that if on the one hand you seek deep transformations in social systems, stimulating sustainable strategies, according to the three pillars of sustainability; on the other hand, there seems to be no concrete efforts to overcome the water crisis.

It is affirmed that there is support for the work "*Manifesto for Democracy on Earth, Justice, Sustainability and Peace*" [2], so the exhaustion of the land, the exploitation of groundwater, the erosion and extinction of biodiversity are succumbed by a lack of governance in the conservation of surface and underground water resources. In addition, in the world scenario, one in three inhabitants suffers from water shortages: for every 1000 liters of water used, another 10,000 are polluted. By 2025, 1.8 billion people are expected to

WIT Transactions on Ecology and the Environment, Vol 234, © 2019 WIT Press
www.witpress.com, ISSN 1743-3541 (on-line)
doi:10.2495/RBM190131

live without water; and only 0.007% of the total water on the planet will be fit for consumption [3].

The connection of these unresolved problems into a larger problem lies in the illusion that nature is self-sufficient, available, abundant; and that water resources, still treated in a peripheral way, evidenced the consequences of their irrational exploitation, in which the implications represent irreversible damages to what should be a vital resource for living beings [4].

In regarding adaptation strategies, it is important to identify and understand production systems that have high resilience to extreme climatic conditions, as these systems can serve as a source of learning and inspiration for the elaboration of strategies to adapt to climate change [5]. If it is no longer possible to prevent climate change from occurring in the next two or three decades, it is still possible to protect our societies and economies from those impacts: for example, by providing quality information, more resilient production systems, improved planning and infrastructure [6]. Therefore, a discussion about the factors or attributes that make groundwater management successful in some regions and not others is the object of this study. This is because it is no longer acceptable for a single organism to decide multiple environmental issues, especially the management and policies surrounding groundwater.

2 A SEMANTIC ISSUE: WATER RESOURCES OR WATER?

The term water serves the natural element, the common good, and is deprived of economic value, just like the waters of rivers, lakes and oceans; whereas, a query of water's character brings a designation of the mercantile use, like the water that is supplied to houses, industries and irrigation for farming. Water as a natural element can be considered free of any use or used, since the water resource is water as an economic unit; however, there is still a conceptual absence between water and water resources; since their legislation makes no distinctions. In addition, water when specifically intended for activities carried out by human beings, becomes a water resource. Added to this concept, the groundwater, found in the saturated zone of the ground, is often denominated like a "poor relative" of the superficial waters [7]. Nevertheless, its operational nature in maintaining ecosystems, for the survival of mankind, accounts for about 97% of the fresh water available on the planet [8]. What refers to the rights to groundwater, has made this natural resource vulnerable to degradation and exhaustion, which occurs mainly due to the deficiency of a peculiar legal regime of management and protection, because these waters constitute the main source of drinking water: for example in Europe, 70% of drinking water and in Brazil 15.6% of the water consumed are from underground sources [9].

In addition, groundwater represents a maintenance function within the hydrological cycle [10]. A fraction of the condensed water returns to the waterways by way of precipitation, or it infiltrates the natural fissures of the soils and rocks, forming the reserves of groundwater that usually return to feed the superficial waters. This balance maintains the renewal of water, and its ecological, biological and thermal functions. The increase in the use of water, in particular underground water, confers human vulnerability onto the balance of the hydrological cycle [11].

It is emphasized that groundwater-free aquifers are those that have a physical connection with surface waters, the holdings are not directly linked to surface waters. This is a criterion that characterizes the aquifers; therefore, indicates the most appropriate management and protection regime. This criterion leads to the understanding that free groundwater is vulnerable to pollution and depletion [10]. This is a fundamental distinction, for a management regime, since this relationship must obligatorily glimpse the relationships between them. With this, information failures, and global and national inequalities that have influenced a new order to reach the whole of society; there is a need for combining decisions and actions to regulate economic activities for the preservation of groundwater [12].

3 THE SUSTAINABILITY DEBATE IN THE SOCIAL CONTEXT AND GOVERNANCE POLICIES

Recognition by a government, or by citizens, of the need for government action is one of the initial steps to create or change a policy. In both political–ideological and practical terms, global governance of development was only born, in fact, a quarter of a century ago, with the rearrangement after the defeat of Nazi fascism. In summary, it is very clear from this rapid overview of the main scientific work that references global ecological boundaries may have been well identified, but that there was no achievement of comparable success in seeking to demarcate them. This fact obviously complicates the noble ambition of not violating these boundary rules, and even more of which environmental changes are really “unacceptable.” Even 20 years after its consecration in Rio de Janeiro, the celebrated sustainable development agenda, which was pompously titled “Agenda 21” has not engendered an even reasonable level of environmental governance [13].

There were countless historical experiences, diplomatic attempts, contemporary movements and actions that have tried to take care of the generation of sustainability, to recover the practices, attitudes and ways of promoting technologies that make us effective and efficient; but to value the role of people in nature, human dignity, empathy, cooperation, equity and commitment. Water, as the essence of sustainability, evaporates, moistens the atmosphere, absorbs ecosystems or covers them with dew. But not only by its watering nature, but because it embodies equity, bathes social justice, penetrates the ecological economy, and involves the human being and the biosphere. This water makes care of nature flourish and mature, brings individual and social transformations, human development; and therefore, universal education, health coverage, culminating in human rights and cultural understanding, which in turn will enable a significant reduction of poverty, real gender equality, coherent eco-technological production, ethically responsible consumption, and attention to diversity and biodiversity. Given this thought, once recognizes the need to create or change a policy that is necessary to effectively implement the management of water resources, in particular groundwater, based on real criteria of accessibility, feasibility and availability, to achieve objectives and goals, adopting in the foreground the change of certain past behaviors [12]. As Morin argues “The techno-economic unification of the globe is in crisis. There is a coincidence between the proliferation of sovereign states, the growth of their interdependence and their ethno-religious closure. This coincidence is not fortuitous. It is explained: a) by the national, ethnic and cultural resistance to Westernization; b) the general fall of the hope placed in Progress. Western civilization itself, which produces the crises of globalization, is in crisis. The selfish effects of individualism destroy old solidarities. A psychic and moral malaise settles at the heart of material well-being. Consumerist intoxications of the middle classes develop, while the situation of the underprivileged classes degrades and inequalities worsen. The crisis of Western modernity makes modernizing remedies for crises oblivious” [14].

The maturity of civilization will undoubtedly be tested by its capacity to overcome the various current environmental crises through concretization and a spirit of cooperation, as well as a sustainable governance system [15]. What is at stake is the way of living hereafter on this planet, within the context of the acceleration of technical–scientific mutations and considerable population growth. There will be no real response to the ecological crisis, except on a planetary scale and on the condition that an authentic political, social and cultural revolution be undertaken, reorienting the objectives of the production of material and immaterial goods. This revolution must, therefore, concern not only the relationships of forces visible on a large scale, but also the molecular domains of sensibility, intelligence and desire. Only a joint action that establishes a collective feeling of affiliation with the entire biosphere will give the opportunity to secure the future. For this, it will be necessary to develop a biospheric consciousness. The empathic civilization begins to emerge, enlarging the whole of a vast project of life that encompasses the planet [16].

The governments are concerned about this, due to the fact that social problems have increased considerably, and consequently environmental problems, and government, as the body responsible for solving them, has not been able to go along hand in hand with the difficulties. The key element in finding solutions to new environmental problems is the promotion of green technologies. For this reason, science and technology occupy a special place in the new environmental policy, necessitating new resources to deal with the ecological crisis. In sum, the perspective of ecological modernization has focused on the new role that science and technology acquire in transforming advanced societies in the face of environmental challenges. There is an impact on the role of science in creating technological innovations that allow prevention and also the solution of environmental problems. It is difficult to imagine a future in which scientific activity plays a central role in the economic, political and cultural terrain of societies. Yet science and technology are not isolated elements, nor the only engine of social change [17].

The way an economy is managed has an impact on the environment; which, in turn, affects both the well-being of the inhabitants and the performance of the economy. Arguing that economic growth leads to environmental degradation and inequalities in wealth, the economy is a subset of our finite environment. As the development of human societies steadily increased and technology has developed increasingly rapidly, so has pressure on the environment. Fueled by an uncontrolled global economic system, which has created so much unprecedented wealth (through excessive consumption) and also huge levels of poverty and environmental problems, these now threaten to increase inequalities and cause irreversible damage to ecosystems on a global scale. Irrevocably, our planet is slowly but surely dying, the ozone layer is running out and global warming has become more evident than ever; leading to drought, famine, flood, misery and disease [18]. There are numerous reasons that cause environmental warming, but the most fundamental cause above all else is the unconsciousness of people about the vital role the environment plays in our life. Undeniably, humans are the most serious culprits of these changes.

Besides the problems of population growth and availability of resources are the issues of environmental depletion, especially of water resources. Unless massive transformations begin toward environmentally sustainable development paths within this decisive decade, environmental deterioration and economic decline will be fueling us, pulling us into social and political decay [14].

Given that there is planetary economic and ecological interdependence, no nation nor income class can isolate itself from the adverse effects of a world experiencing population growth induced by mass poverty, by the destruction of resources driven by excessive consumption and large-scale socio-political disintegration. This obviously also includes corporations, which are totally dependent on the natural world for their resources and livelihood, and on the social world for stability and security.

In the search for alternatives to human development, and thinking about the environmental aspects, one can see the depletion of natural resources, more frequent and intense disasters, loss of biodiversity, environmental degradation, and the negative impact of the increase in global temperature as well as the level of the sea, it makes sense to balance economic development with environmental sustainability and social justice, which is a challenging and complex issue that must be tackled in a systemic way [18]. Managing common equity means making a more integrated use of the human component with the physical component. This necessarily involves a protective attitude towards the environment, which imposes its corrections to the model of sustainable economic development, in order to guarantee the maintenance of the biological balances that are indispensable to the great global balance of the planet. Therefore, it is a citizen conscience directed toward essential attitudes, as a more rational and responsible use of water resources and others from nature, which are not inexhaustible; respect for life in all its forms; reconstruction of that which has been destroyed and adoption of preventive measures [10].

4 ENVIRONMENTAL POLICY OF FREQUENCY GROUNDWATER

The definition of the environmental policy of groundwater is not consensual, because of the breadth of scope of these policies. It covers the complex set of actions and relationships between humans and biodiversity, and all that affects the quality or use of natural resources. As the means to achieve sustainability is premised on modifying personal attitudes and practices; one must allow communities to take care of their own environment, generate a national structure for the integration of development and conservation, and form a global alliance, such as for air and water pollution control [19].

Environmental policies require a determination of the new priorities defined by society, through a new ethic of human behavior and a recovery of social, collective interests, encompassing a set of key changes in the structures of production and consumption, reversing the environmental degradation framework and the social misery it causes, which is not currently occurring. It requires government action focused on new approaches to sustainable development, an objective that requires a balance between "three pillars": the economic, social and environmental dimensions [10]. As well as the capacity and means to effectively combat environmental abuses of either air, soil or waterways.

There is an indispensable need to gather guidelines, principles, strategies, programs and actions to be adopted globally, nationally and locally, and these can be defined as "a planning tool for the construction of sustainable societies", recognizing the "Principle of Supremacy of Interest Public in the Protection of the Environment in Relation to Private Interests". This principle is, in reality, a general principle of modern Public law, by which the superiority of the collective interests is proclaimed, which should prevail over the interests of individuals. It is, in fact, a true assumption of the stability of a social order in which water resources are embedded as a common good [19].

An environmental policy should contribute to minimizing risks and human activities that threaten the health or life of humans or ecosystems; however, it must still be the result of an immense collective effort to transform the environment, which has now become an imaginary war of conquest, dominated by wild nature. Human activities have infinitely surpassed nature's ability to restore its destroyed and consumed elements, and there is unsustainable exploitation of waters [20]. Traditional management aimed at ensuring more rigid regulation has shown that it is impossible to guarantee dignity to people if a healthy environment is not ensured, and it is impossible even to ensure human life without an environment conducive to its development [21]. Alongside social rights, which have been called second-generation rights, today the so-called third-generation rights have emerged, which constitute a category, to tell the truth, still excessively heterogeneous and vague, which prevents us from understanding what effectively can treat the situation. The most important of these is the one claimed by ecological movements: the right to live in an environment that is not polluted [22].

Sustainable development for the protection of groundwater would therefore result in a double pact: a management pact that translates into a constant concern for the conservation and preservation of resources for future generations; and a pact of social organizations expressly concerned about meeting the basic needs of all humans. In addition to social equity and ecological balance, sustainable development presents the third major issue, the issue of economic development. It induces a spirit of common responsibility as a process of change in which the exploitation of material resources, financial investments and technological development routes should acquire harmonious meaning [23]. National environmental policies, as well as international actions for the promotion of sustainability, become public function oriented, for the integrated specialization of the different sectorial policies that intervene in a given territory, and among them, a water resources policy can be executed that considers groundwater as a common heritage, one which means to make integrated use of the human component with the physical component. This necessarily involves a protective attitude of the environment, which imposes corrections to the model of sustainable economic development, in order to guarantee the maintenance of biological balances indispensable to the greater global balance of the planet [20].

With this, a citizen conscience directed to essential attitudes should become established, such as: more rational and responsible use of natural resources that are not inexhaustible; respect for life in all its forms; reconstruction of what has been destroyed; and adoption of preventive measures, including emerging actions, in order to prevent gaps in the use of water resources and ensure the full cooperation of nations, through new and transforming aspects for sustainable development [13]. With this motivation in mind, the challenges to be highlighted lie within the scope of shared action, for the adoption of a shared responsibility, constituting proposals that integrate all the scales, be they municipal, national or global, of environmental policies that are more effective at protection of groundwater; as well as the implementation of a global political consensus, providing a focus for advocacy, improving the targeting and flow of aid, as well as the follow-up of development projects that provide a reference to address the most pressing social challenges in the world, with a priority focus on meeting basic human needs, access to and availability of water use in a truly sustainable way [23]. In this argument, groundwater protection policies can help public understanding of the complex challenges for sustainable development, inspiring public and private action, promoting integrated thinking and encouraging accountability. The challenges addressed are inherently integrated, so sustainable development will require actions together rather than individually or one at a time. Thus, efforts will be necessary so that social participation does not end up being only a form of bureaucratic legitimization of the decisions that were previously taken.

5 FINAL CONSIDERATIONS

Debates on issues related to underground water availability as a source of water supply have raised major concerns about the quantity, quality and depletion of this resource. In this context, the greatest challenge in terms of water resource conservation lies in the lack of effective mechanisms to control, protect and use these natural resources, without impeding the economic growth of the country. On the other hand, the concept of a water crisis goes beyond scarcity, the result of a combination of failures of governance, environmental preservation, sustainability and disparity in access to water. The different countries will have to present different degrees of attention and effort to incorporate the human right to water; guaranteeing safe access, replenishment of the aquifer's storage capacity, and the need for a more active role of sustainability in society, requiring actions of rationality, as important in resource allocation as in equity, which is often a crucial element in governance decisions.

Therefore, implementing effective water resource management policies is not a common process, since it establishes the participation of values and that depend fundamentally on the resources made available by nature, on the maintenance and sustainability of water resources, aiming at collective sustainability in order to create legal and public policies of sustainable development, allied to the system of government founded constitutionally in accordance with the dictates of social justice, politics and health.

REFERENCES

[1] Bauman, Z., *Globalização: As Consequências Humanas* (*Globalization: Human Consequences*), trans. M. Penchel, Zahar: Rio de Janeiro, p. 32, 1999.

[2] Shiva, V., *Manifiesto para una Democracia de la Tierra. Justicia, Sostenibilidad y Paz* (*Manifesto for Democracy on Earth, Justice, Sustainability and Peace*), Paidós: Barcelona, pp. 221–217, 2006.

[3] Associação Brasileira de Águas Subterrâneas (ABAS), *Águas subterrâneas* (*Groundwater*), Campinas, SP, 2007.

[4] Amaral, S. et al., Comunidades ribeirinhas como forma socioespacial de expressão urbana na Amazônia: uma tipologia para a região do Baixo Tapajós (Pará-Brasil) (Riverine communities as a socio-spatial form of urban expression in the Amazon: a typology for the region of the lower Tapajós (Pará-Brazil)). *Rev. Bras. Est. Pop.*, **30**(2), pp. 367–399, 2013.

[5] de Santos, B. & Garavito, C.A.R. (eds), *El Derecho y la Globalización desde Abajo: Hacia una Legalidad Cosmopolita* (*Rights and Globalization from below: Towards Cosmopolitan Legality*), Anthropos: Barcelona, p. 340, 2007.

[6] Irigaray, C.T.H., Água: um direito fundamental ou uma mercadoria? (Water: A fundamental right or a commodity?) *Congresso Internacional de Direito Ambiental: Direito,* ed A.H. Benjamin, Água e Vida, Imprensa Oficial: São Paulo, p. 118, 2003.

[7] McCaffrey, S.C., The contribution of the UN Convention on the law of the nonnavigational uses of international watercourses. *International Journal of Global Environmental Issues*, **4**(3), pp. 250–263, 2001.

[8] Salif-Diop, P.R., *Atlas Mondial de l'Eau: Une Pénurie Annoncée* (*World Atlas of Water*), Autrement/PNUE/Mémorial de Caen: Paris, p. 16, 2003.

[9] Chazournes, L.B., Eaux internationales et droit international: vers l'idée de gestion commune, Recueil des Cours de l'Academie de la Haye, p. 34, 2005.

[10] Granziera, M.L.M., *Direito de Águas: Disciplina Jurídica das Águas Doces* (*Water Law: Legal Discipline of Sweet Water*), Atlas: São Paulo, p. 64, 2001.

[11] Machado, P.A.L., *Recursos Hídricos. Direito Brasileiro e Internacional* (*Water Resources. Brazilian and International Law*), Malheiros: São Paulo, p. 55, 2002.

[12] Estevan, A.E. & Naredo, J.M., *Ideas y Propuestas para una Nueva Política del Agua en España*, Fundación Nueva Cultura del Água: Bilbao, p. 119, 2004.

[13] Veiga, J.E., Indicadores de sustentabilidade (Sustainability indicators). *Estudos Avançados*, **24**(38), pp. 39–52 and p. 112, 2010.

[14] Morin, E., *A Via para o Futuro da Humanidade* (*The Way to the Future of Humanity*), 2nd ed., Tradução de Edgard de Assis Carvalho, Mariza Perassi Bosco, Bertrand Brasil: Rio de Janeiro, pp. 23–26, 2015.

[15] Berggruen, N. & Gardels, N., *Governança Inteligente para o Século XXI* (*Intelligent Governance for the 21st Century*), Prisa Edições: Rio de Janeiro, p. 236, 2013.

[16] Rifkin, J., *La Civilización Empática* (*Empathetic Civilization*), Paidós: Barcelona, p. 606, 2010.

[17] Beck, U., *La Sociedad del Riesgo: Hacia una Nueva Modernidad* (*The Society of Risk: Towards a New Modernism*), Ediciones Paidós: Barcelona, pp. 20–12, 1998.

[18] Riechmann, J., Reyes, L.G., Herrero, Y. & Madorrán, C., *Qué Hacemos Hoy cuando nos Encontramos Frente a la Amenaza de una Crisis Mayor que la Económica: La Ecológica* (*What to do Today when Faced with the Threat of a Crisis Larger than Economic: Ecologic*), Ediciones Akal: Madrid, pp. 10–11, 2012.

[19] Dye, T.R., *Understanding Public Policy*, Prentice-Hall: Upper Saddle River, p. 107, 2002.

[20] Jacobi, P.R. & Barbi, F., Democracia e participação na gestão dos recursos hídricos no Brasil. *Katálysis:Florianópolis*, **10**(2), pp. 237–244, 2007.

[21] Jorgensen, S.E. & Rast, W., The use of models for synthesizing knowledge for integrated lake basin management, and facilitating implementation of the World Lake Vision. *Lakes and Reservoirs*, **12**, pp. 3–13, 2007.

[22] Dasgupta, P., The idea of sustainable development. *Sustainability Science*, **2**, pp. 5–11, 2007.

[23] Varela, C.A., Instrumentos de políticas ambientais, casos de aplicação e seus impactos para as empresas e a sociedade. *Revista Ciências Administrativas Fortaleza*, **14**(2), pp. 251–262, 2008.

ASSESSMENT OF THE ECOLOGICAL POTENTIAL OF HEAVILY MODIFIED WATER BODIES IN THE BASILICATA REGION, ITALY: CASE STUDY OF THE AGRI RIVER BASIN

FRANCESCO ARBIA[1], LORENA DI GIUSEPPE[1], RAFFAELE GIAMPIETRO[1], MICHELE GRECO[1,2*], PATRIZIA MAURO[1], GIOVANNI MARTINO[2], ANNA MONTELLA[1], GIOVANNI MUSSUTO[1], ERIKA NAPOLI[1] & ANTONELLA SMALDONE[1]
[1]Regional Environmental Research Foundation – FARBAS, Potenza, Italy
[2]School of Engineering, University of Basilicata, Italy

ABSTRACT
River morphology and ecosystem are macro-indicators of climate and land-use changes as well as human activities and land policies. In the last decades, knowledge of water resources, in terms of quality, quantity, prevention and control, has quickly increased – even for the availability of advanced methodologies and technologies both for monitoring and modelling fields. Nevertheless, the objective of a good ecological status of water bodies has not yet been reached due to human pressure which still induces long term impacts as well as direct or indirect physical constraints. In such a context, the European Water Framework Directive introduces the concepts of artificial water bodies (AWB) and heavily modified water bodies (HMWB) and states that in such cases good ecological potential must be reached while good ecological status is required only for natural water bodies (NWB). Further, the methodologies proposed for the AWB and HMWB need a wide set of parameters and measurements, which are not always systematically available, inducing further actions and analyses in order to evaluate the quality level of the water bodies. Moreover, the AWB or HMWB classification represents an opportunity to resize the measures addressed to achieve good ecological status both in terms of economic engagement and timing with relevant consequences on water resource planning and management. The paper deals with these issues and synthetically reports the work done on the whole Lucanian river network (Basilicata region, Southern Italy) and also presents the Agri River case study in order to classify, as best as possible, AWB, HMWB and NWB as well as to support survey planning.
Keywords: water framework directive, good ecological potential, good ecological status, heavily modified water bodies, artificial water bodies, natural water bodies.

1 INTRODUCTION

The main goal of the European Water Framework Directive issued by the European Commission [1] is to ensure water quality protection and improvement throughout the European Union (EU) by enforcing a common water resource management and control policy as well as by defining powerful strategies aimed to reduce pollution and natural resource overexploitation [2]. Such a Directive was promulgated in 2000 and defines the deadlines to reach the objective of a good ecological level of integrity for all the types of water bodies: rivers, lakes, transitional water, coastal water and groundwater [3]. In detail, the European water policy is addressed to achieve the strategic objective of good chemical status (GCS) as well as good ecological status (GES) of natural water bodies (NWB) and groundwater bodies (GWB). Otherwise, EU policy aims to reach good ecological potential (GEP) for the defined heavily modified water bodies (HMWB) and artificial water bodies (AWB) by 2015 or 2021

**ORCID: http://orcid.org/0000-0002-3986-7117*

www.witpress.com, ISSN 1743-3541 (on-line)
doi:10.2495/RBM190141

or even 2027 under specific conditions. That is, in order to reach such a target, operative tools of integrated river basin management process have been created and promoted [4], [5].

In such a framework two main relevant aspects concerning the classification of NWB, HMWB as well as AWB arise. In fact, for NWB the GES is related to ensure only slight measures or actions in order to reach 'completely' or 'almost completely' undisturbed conditions. Otherwise, HMWB and/or AWB conditions might be identified and designated [6] before defining further restrictive measures. According to EC-WFD [1], hydro-morphological conditions [7] as well as the uses of the water resources represent the basis to appoint water bodies as HMWB.

The water body designation as artificial or heavily modified occurs when, amongst other things, changes in the hydro-morphological characteristics would have significant adverse effects, for example, on flooding, navigation, hydro-power plants and so on. The target for such water bodies is to reach GEP, which corresponds to reduced environmental objectives [1], [8]. Moreover, the EC-WFD requires a severe reduction and prevention of the water quality deterioration for NWB, AWB and HMWB.

Furthermore, the classification must follow an intensive and well-structured monitoring plan requiring both chemical–physical and biological surveying in order to obtain an exhaustive framework concerning the effects and the impacts induced by natural and human activities.

The paper proposes the methodology applied in Basilicata (Southern Italy) and finalised to evaluate a possible classification as HMWB for several surface reaches in order to establish the real and coherent criticality scenario. In fact, the Basilicata region has six main rivers which were strongly affected by the construction of several reservoirs for agricultural and drinkwater purposes in the sixties. Hence, several dams and weirs were built during those decades, in which no environmental regulations were adopted according to the general global technical and social-economic culture. Of course, no legal actions can be applied at present, but because long-term effects still persist, structural interventions for hydro-morphological restoration in order to reach GEP are required. On the other hand, the age of this hydraulic work induces us to consider water bodies as natural because infrastructures are well integrated in the environmental landscape as well as a biodiversity producer. Furthermore, the general feeling of the local people about a dam built up in the sixties is to consider it as a lake with all the positive and negative related feedback. Besides, the upstream and downstream water body characteristics are still influenced by the dam/reservoir both in terms of hydraulics/hydrology and morphology as well as sedimentary dynamics. In such a case, even if the water discharge changes, the evaluation can always be performed by direct measurement or expeditive assessment [9], [10] and sustainable remediation system for sediment continuity might be proposed [11]. The local natural environment is compromised and only the concept of GEP can be reached.

2 STUDY AREA – THE AGRI RIVER BASIN

The proposed analysis has been applied to the whole of Lucanian river networks (Basilicata region, Southern Italy) but the present study refers only to the Agri River as a suitable example representing all the problems observed on the Lucanian river system. In detail, the study area consists of the Agri river basin. Agri river is one of the six major rivers of the Basilicata region, with a drainage basin area over 1715 km^2 and varying morphology from mountainous and hilly, in the medium-high upstream part, to low hilly and flat, in the downstream portion. The hydrographic network is substantially ramified, presenting a main stream of about 113 km long, whose mountainous reach with NNW–SSE trend, crossing the intermontane depression of the Alta Val d'Agri and then assuming a fairly regular W–E trend reaching the Ionian coast of Lucania.

The average annual rainfall is quite homogeneous along the basin, following the distributions aligned to the NW–SE ridge with a reference value of 700–900 mm/yr.

From a geological point of view, the Agri River basin is characterized by a complex geological-structural order (Fig. 1) and, for this reason, the hydrogeology of a large part of the territory in that area is very complex (Fig. 2).

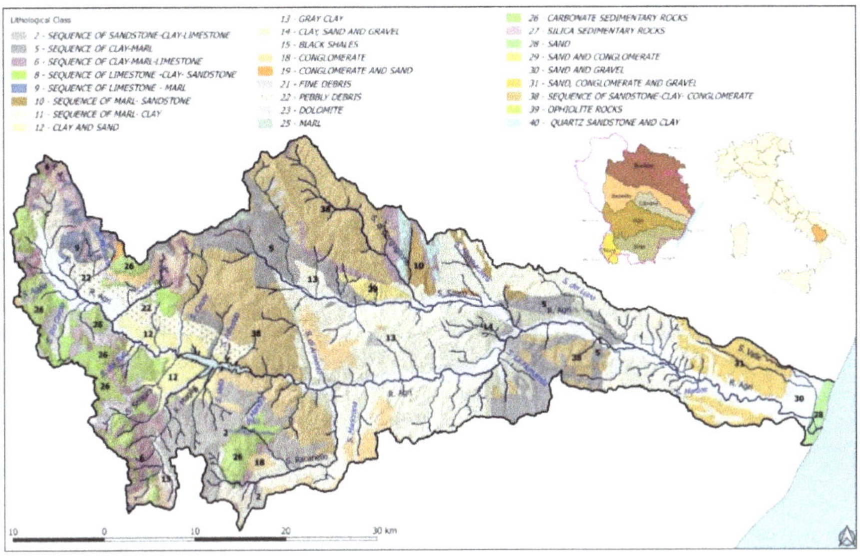

Figure 1: Geo-lithological classes and river network of Agri basin.

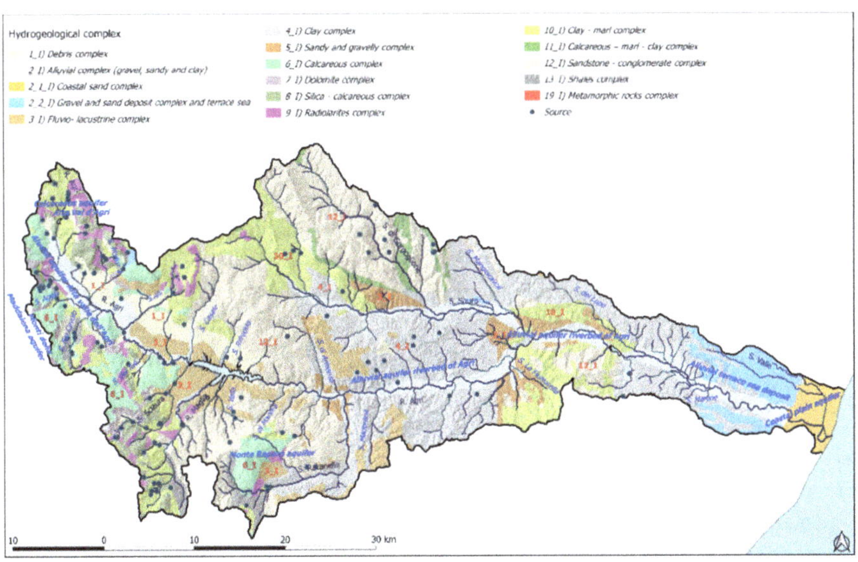

Figure 2: Hydrogeological system of the Agri basin.

The carbonate rocks characterizing the main mountain ranges of the Alta Val d'Agri, in tectonic contact with clayey-marly units and flyschoids, represent the strategically most important aquifers in the basin.

In addition to the aforementioned hydrogeological domains, there are others that are less significant from the point of view of their water potential, but with a considerable interest and value. More specifically, it is worth highlighting:

a) the sandy-conglomeratic aquifers falling in the middle and lower portions of the basin;
b) the sandy and gravelly alluvial aquifers present in the down stream valley;
c) the sandy aquifer falling into the Coastal Plain.

Furthermore, the Alta Val d'Agri area is characterized by calcareous and calcareous-dolomitic succession mesozoic platforms, which occupy a large part of the north-western basin and by meso-cenozoic succession of deep basin, originating from the deformation of a dominion oceanic tetidus, and from Lagonegro-type basin units, covered in discordance by myocenic synoptic clastic formations.

Moreover, most of the valley area is characterized by deposits of fluvio-lacustrine basin generated by a system of plio-quaternary faults having preferential directions N 120° ± 10°. Recent studies highlight a transversal structure of the basin similar to an irregular graben, asymmetrical towards the north-eastern margin.

Longitudinally, the basin consists of three main depocenters separated by two structural high-oriented NE–SW. The meso-cenozoic tectonic units, present in this area, are attributed to different original contiguous paleogeographic domains.

3 MORPHOLOGICAL QUALITY INDEX FOR THE AGRI RIVER

In Italy, the EC-WFD has been adopted through the introduction of Morphological Quality Index method (MQI), described in detail in reference guide IDRAIM [12]. It is quite important to point out that the Italian legislative system adopts MQI as an official methodology for the classification of the biological, chemical and hydro-morphological status of water bodies.

The methodology is based on evaluating the deviation from a given reference morphological state of each water body. Further, the approach takes into account the temporal context, in terms of trends of 'trajectory' and morphology of the fluvial system in order to understand its current conditions.

The first phase of MQI evaluation procedure provides a general setting of the physical conditions of the river and carries on an initial classification in relatively homogeneous reaches based on physiography, channel confinement, channel pattern, river typology and artificial interventions.

Further, the second phase consists of real measurement of MQI for all of the homogeneous selected reaches, in order to classify the present morphological state according to three different components:

a) morphological functionality: based on the observation of processes in the present conditions;
b) artificiality: in which frequency and impact of artificial interventions are evaluated;
c) morphological channel changes: recent morphological variations (referring to the planimetric changes in the last 50–60 years) are examined.

These parameters of functionality, artificiality and morphological channel changes are investigated by using specific assessment procedures, based on schematic reports, which allow us to analyse the different conditions constraining the status of the stream.

For this reason, a number of indicators are used, providing both qualitative and quantitative evaluation of variables and schematic reports are differentiated for various aspects, depending on river typologies (confined channels or/and semi-unconfined channels) in order to provide an exhaustive morphological characteristic value. However, the MQI methodology is not suitable for assessing changes in morphological quality of small channels, and it is not applicable to artificial water bodies, lakes, or reservoirs.

The main idea deals with the measurement of the deviation between the morphological conditions of the considered stream and undisturbed conditions which represent the optimal status. In detail, the following reference classes are considered:

- class A: undisturbed conditions or negligible alterations score 0;
- class B: intermediate alterations scores from 2 to 5;
- class C: very altered conditions scores from 6 to 12.

Thus, the total deviation S_{tot} from not-altered conditions is calculated as the sum of the scores assigned to all indicators and it provides an assessment of the deterioration of the reaches compared to their natural conditions.

This value can be normalized with respect to S_{max}, that is the maximum score that could be obtained once all indicators assume the maximum possible score. Further, the Morphological Alteration Index (MAI) represents the normalized assessment of the morphological alteration of the considered water body and is obtained by the ratio [12]:

$$MAI = S_{tot}/S_{max} \tag{1}$$

The Morphological Quality Index (MQI) is then defined as [12]:

$$MQI = 1 - MAI \tag{2}$$

Such an index assumes value 1 in the case of reference conditions (corresponding to totally natural condition) and value 0 in case of maximum alteration of the water body. All other conditions fall in between these limits. Moreover, different classes of morphological quality are defined as reported in Table 1.

Table 1: Morphological Quality Index classification.

MQI	0.0≤MQI<0.3	0.3≤MQI<0.5	0.5≤MQI<0.7	0.7≤MQI<0.85	0.85≤MQI<1.0
Quality class	Bad status	Poor status	Moderate status	Good status	High status

Fig. 3 shows the resulting MQI classification for the Agri River drainage network referring to all the water bodies selected in the basin and codified through the Water Information System for Europe (https://water.europa.eu/). The data used for the evaluation refer to an intensive measurement campaign carried out in 2016–2017 by the Regional Environmental Protection Agency (ARPAB).

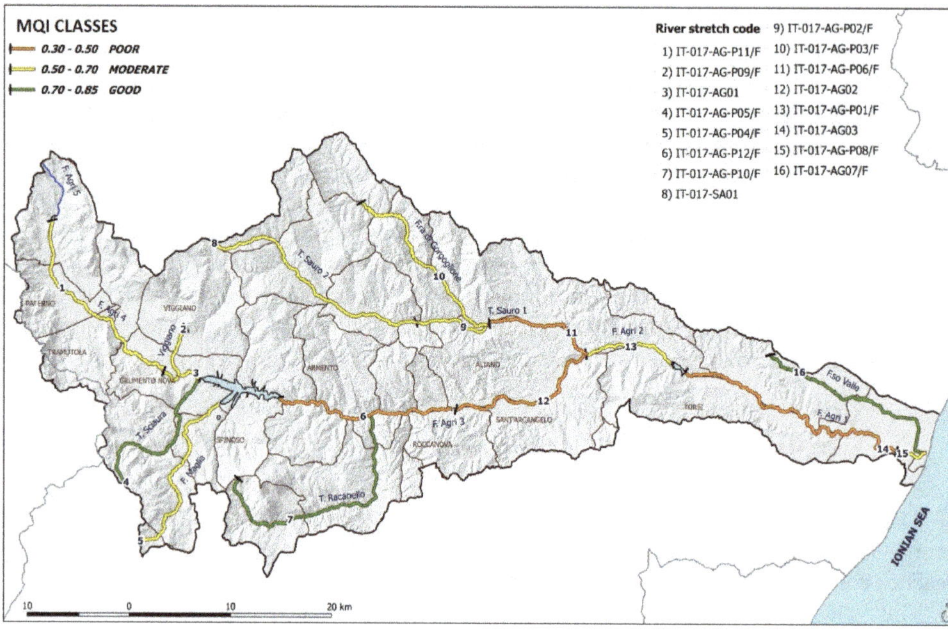

Figure 3: Morphological Quality Index classification for the Agri River basin.

4 ASSESSMENT OF ECOLOGICAL POTENTIAL OF HMWB

The methodology, described in the IDRAIM manual [12], is divided into two phases and reflects the contents of the European Guidance document [6]:

a) 'PHASE 1' is defined as the 'Preliminary Identification' procedure: this phase involves the identification of water bodies that provisionally could be indicated as HMWB basing on technical, hydro-morphological and ecological evaluations;
b) 'PHASE 2' is defined as the 'Designation' procedure: in this phase the HMWB are definitively designated and the effective recognition requires both technical and socio-economic assessments and complex political choices.

In detail, the morphologically homogeneous reaches, characterized by high and significant hydro-morphological alterations, are preliminarily identified as HMWB, through cases of attribution, defined by the National Institute for Environmental Protection and Research (ISPRA) [12] and listed in Table 2. For each considered water body, accurate analyses are performed in order to verify the occurrence of at least one of the listed significant hydro-morphological alteration cases.

In the case of Agri River, 16 water bodies have been assumed as representative of the whole basin and have been surveyed and monitored performing biological and chemical analyses (Fig. 4) providing a detailed monography of the main stream and all possible significant hydro-morphological alterations affecting the system.

The general and clear framework of the system of water bodies belonging to the Agri River network is obtained by combining the sets of biological and chemical analyses and monitoring, with the results of surveys addressed to reveal high and significant hydro-

Table 2: List of reference for high and significant hydro-morphological alteration cases.

1	Transversal structures (including bed sills and ramps) with density >1 every *n*, where *n* = 100 m in mountain landscape and *n* = 500 m in flat/hilly area
2	Bank protection and/or levees at contact with most of the water body (> 66%)
3	Presence of bed revetments for most of the water body
4	Presence of a dam (or check dams and weirs) at the upstream boundary of the water body
5	Presence of transversal structures (check dams or weirs) inside the water body or at the downstream boundary which cause high alteration of hydrodynamic conditions and artificial lentic reaches
6	High alteration of hydrodynamic conditions in the reaches (significant decrease or increase of flow)
7	Alteration of hydrodynamic characteristic of water body due to hydropeaking
8	Combination of several permanent pressures, none of which are included in the cases 1–7, but the combination results in a significant alteration of the water body

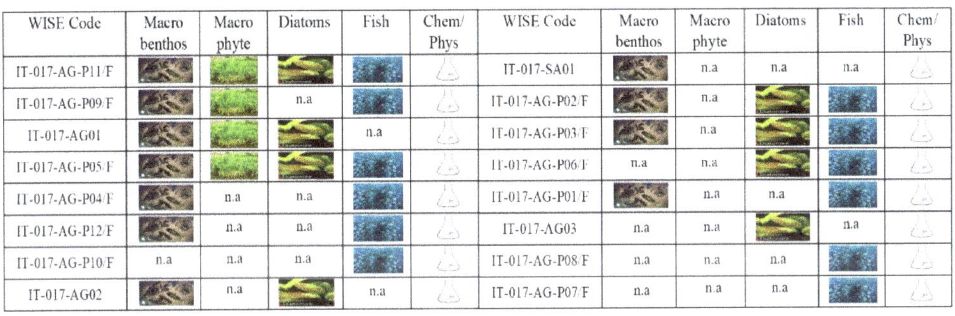

WISE Code	Macro benthos	Macro phyte	Diatoms	Fish	Chem/ Phys	WISE Code	Macro benthos	Macro phyte	Diatoms	Fish	Chem/ Phys
IT-017-AG-P11/F						IT-017-SA01		n.a	n.a	n.a	
IT-017-AG-P09/F			n.a			IT-017-AG-P02/F		n.a			
IT-017-AG01				n.a		IT-017-AG-P03/F		n.a			
IT-017-AG-P05/F						IT-017-AG-P06/F	n.a	n.a			
IT-017-AG-P04/F		n.a	n.a			IT-017-AG-P01/F		n.a	n.a		
IT-017-AG-P12/F		n.a	n.a			IT-017-AG03	n.a	n.a		n.a	
IT-017-AG-P10/F	n.a	n.a	n.a			IT-017-AG-P08/F	n.a	n.a	n.a		
IT-017-AG02		n.a		n.a		IT-017-AG-P07/F	n.a	n.a	n.a		

Figure 4: Framework of biological and chemical analyses performed on the water bodies.

morphological alterations. Figs 5 and 6 show the ecological status and the ecological potential maps for the Agri River respectively. Table 3 reports a comparison outlining in which water body the high and significant hydro-morphological alterations occur and the relative ecological status, as inferred through the bio-chemical analyses and the ecological potential as well.

The operative results are quite interesting, showing an overall positive picture of the Agri River system. In fact, the assessment of the ecological status of water bodies seems to provide a crucial and critical scenario, due to the relevant percentage of water bodies in moderate ecological status that is close to 40% of the total water bodies. On the other hand, taking into account the possibility to classify some of the water bodies as HMWB, the total amount of water bodies in moderate ecological status sensitively decreases to 16% and the general status increases the quality triggering the good/high ecological potential.

Such results have an interesting and relevant feedback on the basin action plan addressing the measures and interventions to be implemented in order to reach the Good Ecological Status (GES) for those low-quality water bodies. That is, the Agri River has a very good ecological potential with a positive feedback for the whole system. Furthermore, the quantitative effect of the aforementioned results deals with the effectiveness of the

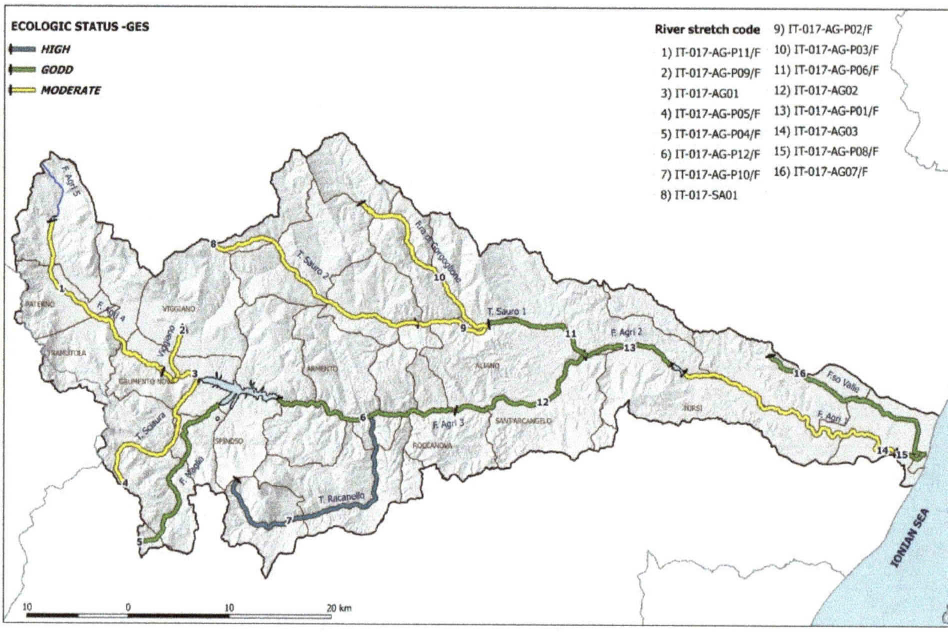

Figure 5: Ecological status of Agri River water bodies.

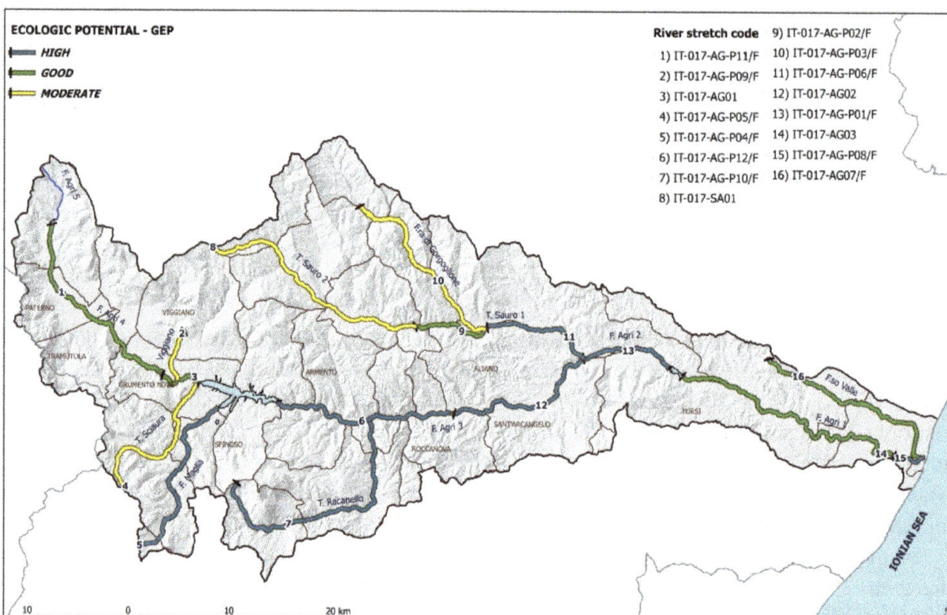

Figure 6: Ecological potential of Agri River water bodies.

Table 3: General framework comparing the ecological status, provisional identification of HMWB and the ecological potential for the Agri river system.

WISE code	Ecological status	Provisional identified HMWB	Ecological potential
IT-017-AG-P11/F	MODERATE	**HMWB**	GOOD
IT-017-AG-P09/F	MODERATE	NO HMWB	MODERATE
IT-017-AG01	MODERATE	**HMWB**	GOOD
IT-017-AG-P05/F	MODERATE	NO HMWB	MODERATE
IT-017-AG-P04/F	GOOD	**HMWB**	HIGH
IT-017-AG-P12/F	GOOD	**HMWB**	HIGH
IT-017-AG-P10/F	HIGH	NO HMWB	HIGH
IT-017-AG02	GOOD	**HMWB**	HIGH
IT-017-SA01	MODERATE	NO HMWB	MODERATE
IT-017-AG-P02/F	MODERATE	**HMWB**	GOOD
IT-017-AG-P03/F	MODERATE	NO HMWB	MODERATE
IT-017-AG-P06/F	GOOD	**HMWB**	HIGH
IT-017-AG-P01/F	GOOD	**HMWB**	HIGH
IT-017-AG03	MODERATE	**HMWB**	GOOD
IT_017_AG_P08/F	GOOD	**HMWB**	HIGH
IT-017-AG-P07/F	GOOD	NO HMWB	GOOD

sustainable policy developed in the last few years on the water resources and the relevant reduction of the amount of investments to be employed to resolve problems as well as to maintain such a good status of the river system. That is, the AWB or HMWB classification represents an operative tool to review the measures addressed to achieve the good ecological status in terms of economic engagement and scheduling augmenting the positive effects on water resource planning and management.

The extension of such approach to all the Lucanian rivers allow us to define a good scenario supporting the related water resource policies to be implemented for maintaining and further increasing the ecological status.

5 CONCLUSIONS

The macro-components of river morphology and ecological system are sensitive indicators of natural and human alterations due to climate and land-use changes and management as well as territorial policies.

The European Water Framework Directive establishes common policies in the field of water resource management and control able to ensure the objective of water quality protection in the EU Countries because water is a heritage which must be protected, defended and treated as such. Thus, the Directive aims to achieve the goal of a good ecological level of integrity even for fresh water, groundwater, transitional water and marine water as well.

Further, the WFD introduces the concepts of artificial water bodies (AWB) and heavily modified water bodies (HMWB) and states that in such cases good ecological potential (GEP) must be reached while good ecological status (GES) is required only for natural water bodies (NWB). Moreover, the increasing knowledge in advanced methodologies and technologies for both monitoring and modelling has strongly augmented the institutional capability for water resources management, in terms of quality, quantity, prevention and control. In such a context, the paper deals with the operative implementation of the methodology adopted for the Lucanian rivers and describes the Agri River case study in order to propose the current surface water classification using heavily modified water bodies as reference.

The obtained results show an overall positive vision of the river drainage network which rates from moderate level to high level of ecological potential. Hence, such results highlight not only the effectiveness of the employed local water resource policies but very interesting prospective effects on the forthcoming regional water basin management aimed to completely ensure EC-WFD objectives.

Finally, the AWB or HMWB classification represents an opportunity to resize the measures addressed to achieve the good ecological status both in terms of economic engagement and timing with relevant effects on water resource planning and management.

ACKNOWLEDGEMENTS

This work has been granted by Regione Basilicata – DGR 1490/2014 – Project "*INDICARE – Indicizzazione delle criticità ambientali regionali della Basilicata*" and carried out by the Regional Environmental Research Foundation (FARBAS).

REFERENCES

[1] EC-WFD, Directive 2000/60/EC of the European Parliament and of the Council of 23 October 2000 establishing a framework for Community action in the field of water policy. European Water Framework Directive. *Official Journal of the European Communities*, L 327/1, 2000.

[2] Koenig, F., Quick, I. & Vollmer S., Defining quantitative morphological changes in large rivers for a sustainable and effective sediment management applied to the River Elbe, Germany. *Proceedings of the Tenth International Conference on Hydroscience & Engineering*, pp. 4–8, 2012.

[3] Cron, N., Quick, I. & Zumbroich, T., Assessing and predicting the hydromorphological and ecological quality of federal waterways in Germany: development of a methodological framework. *Hydrobiologia*, 2015. DOI: 10.1007/s10750-015-2484-x 2015.

[4] Weiß, A., Matouskova, M. & Matschullat, J., Hydromorphological assessment within the EU-Water Framework Directive—trans-boundary cooperation and application to different water basins. *Hydrobiologia*, **603**(1), pp. 53–72, 2008.

[5] Raven, P.J., Holmes, N.T.H., Charrier, P., Dawson, F.H., Naura, M. & Boon, P.J., Towards a harmonized approach for hydromorphological assessment of rivers in Europe: a qualitative comparison of three survey methods. *Aquatic Conservation: Marine and Freshwater Ecosystems*, **12**(4), pp. 405–424, 2002.

[6] CIS, Guidance document no. 4 – Identification and Designation of Heavily Modified and Artificial Water Bodies. Produced by working Group 2.2 – HMWB. CIS – Common Implementation Strategy for the Water Framework Directive (2000/60/EC), Luxemburg, 2003.

[7] Benjankar, R., Koenig, F. & Tonina, D., Comparison of hydromorphological assessment methods: application to the Boise River, USA. *Journal of Hydrology*, **492**, 128–138, 2013.

[8] CIS, Guidance document n. 13 – Overall Approach to the Classification of Ecological Status and Ecological Potential. Produced by working Group 2A. CIS – Common Implementation Strategy for the Water Framework Directive (2000/60/EC), Luxemburg, 2005.

[9] Mirauda, D., Greco. M. & Moscarelli, P., Practical method for flow velocity measurements in fluvial sections. *WIT Transactions on Ecology and the Environment,* **146**, pp. 355–366, 2011.
[10] Greco, M. & Martino, G., 1-D versus 2-D entropy velocity law for water discharge assessment in a rough ditch. *Entropy 2018*, **20**(9), p. 638, 2018.
[11] Carone, M.T., Greco, M. & Molino, B., A sediment-filter ecosystem for reservoir rehabilitation. *Ecological Engineering*, **26**(2), pp. 182–189, 2006.
[12] Rinaldi M., Surian N., Comiti F. & Bussettini M., *IDRAIM, Sistema di valutazione idromorfologica, analisi e monitoraggio dei corsi d'acqua*, ISPRA, Manuali e Linee Guida 113/2014, Roma, 2014.

DEVELOPMENT OF A RAINFALL-RUNOFF MODEL: SUITABILITY ASSESSMENT IN THE MORAVA RIVER BASIN, CZECH REPUBLIC

KATERINA KNOPPOVA[1,2] & DANIEL MARTON[1]
[1]Institute of Landscape Water Management, Brno University of Technology, Czech Republic
[2]Czech Hydrometeorological Institute, Czech Republic

ABSTRACT

Availability of water as a natural resource is affected by the changing climate. To assess local hydrological impacts of climate change, less or more complicated hydrological models are being used. This paper presents a novel rainfall-runoff software, Runoff Prophet, which can simulate monthly discharge in a confluence of any river catchment. Usage of the model is based on monthly climatic and hydrological data as an input with no need of catchment characteristics description such as river slope, catchment slopes, land cover, river net system etc. Model calibration is based on differential evolution optimization algorithm, Thornthwaite evapotranspiration estimation method and set of water balance equations. The developed software was tested in complicated hydrological conditions in the Czech Republic. To assess universality of the newly developed software, model calibration and validation was performed and evaluated for catchments of different types across the Morava River Basin. The applicability of the software was tested for representative catchments classified into five pre-determined categories: i) agricultural, ii) forest, iii) mountain, iv) urbanised, and v) combined land use river basin. It was confirmed that Runoff Prophet is an effective tool for rainfall-runoff modelling with the purpose of long-term projection. In practice, it should be used for predictions of changes in hydrological balance of landscape or for estimations of future water supplies in reservoirs when assessing the current ones, as well as designing new ones.

Keywords: Runoff Prophet, rainfall-runoff model, lumped, hydrological modelling, Nash–Sutcliffe.

1 INTRODUCTION

In recent years, the impacts of climate change are being felt strongly in the Czech Republic. In future, these changes are expected to continue in the same extent and probably even deepen [1]. According to the Czech Hydrometeorological Institute preliminary press release [2], our country is still suffering from a dry period lasting since 2014. One of the indicators of drought is a low value of mean annual rainfall in comparison to long-term mean annual precipitation sum, which is 686 mm in the Czech Republic for the reference period 1981–2010. The most critical year in this dry period was 2015 with 187 mm mean annual rainfall deficit (28%), and the year 2018 was evaluated as the second driest year with 171 mm deficit (25%). With regards to the hydrological impact in the form of low river flows, the situation in the Czechia was worst in 2018 due to long-term accumulation of water deficit from previous years. For the Czech policymakers it is crucial to know whether the dry period is just a rare extremity or if one should expect serious problems due to lack of water in the future.

It is possible to make projections of future hydrological situation based on the available climate data predictions, provided that one is able to sufficiently describe the rainfall-runoff process in the particular catchment. The relationship between precipitation amount and river discharge is influenced by many factors, such as the basin area size and topography, hydrographic network, geological and soil parameters, land cover or the nature of the rainfall, temperature and soil moisture. It is a complex natural process and it is not a simple task to find clear relationships between climate data and discharges.

WIT Transactions on Ecology and the Environment, Vol 234,
www.witpress.com, ISSN 1743-3541 (on-line)
doi:10.2495/RBM190151

Rainfall-runoff models are tools enabling simulation of this natural cycle. Many of them are based on description of the catchment characteristics, which tends to be one of the most challenging and time-consuming steps of hydrological modelling. This includes for example the models HEC-HMS [3], SAC-SMA [4], HYDROG [5], SWAT [6], TOPMODEL [7] and many others. The models are often locked without the possibility of advanced user custom modifications. They are quite complex, with strict requirements on input data quantity. However, it is possible to simulate rainfall-runoff process in an easier way, without having the catchment description available. Conceptual and empirical models are based on this counter-attitude. The relationship between cause and consequence is searched using calibration datasets of input elements. One of such models is also the Runoff Prophet [8], which was developed at the Institute of Landscape Water Management at Brno University of Technology.

The model is based on water balance equations used by Wang et al. [9] for large catchments in China, where the transition between a wet and dry season is quite evident. The original equations were successfully transferred to moderate climate conditions of the Czech Republic [10]–[12], where the hydrological regime significantly differs by its variability. This implies that the equations, and thus Runoff Prophet, could be transferable to similar basins in Temperate climate zones, and under certain circumstances worldwide.

Runoff Prophet was successfully tested on three case studies, but a systematic evaluation of the model efficiency has not yet been made. The aim of this paper is to examine the universality of the developed model and the equations used; and to find out whether it is applicable for catchments of different types across the Czech Republic natural conditions. The purpose of this paper is to determine whether the accuracy of the rainfall-runoff process simulation is affected by land use type of the river basin. The paper also attempts to ascertain the connections between model efficiency and the catchment area, number of available weather stations with satisfactory data and other individual catchment characteristics. The results of Runoff Prophet calibration and validation for 15 catchments classified into five representative land use types are presented.

2 METHODS

The basis of the developed hydrological model are the water balance equations used by Wang et al. [9]. They are intended for monthly calculation time step and cover two main components of total discharge from the basin: surface flow and groundwater flow. They also take into account potential evapotranspiration and soil moisture. In the basic form, they include four unknown variables: initial soil moisture S_0, coefficient of surface flow k_s, coefficient of groundwater flow k_g and coefficient of evapotranspiration k_e.

For Runoff Prophet, monthly variability of calibration coefficients was selected. Thus there are 37 variables to be calibrated – S_0, 12 $k_{s,i}$, 12 $k_{g,i}$ and 12 $k_{e,i}$, where $i = 1, 2, \ldots, 12$ (January to December). To evaluate the chosen set of variables, the Nash–Sutcliffe model efficiency coefficient (NS) was used as a fitness function to be maximized. Model calibration is based on the differential evolution described in [13]. Evapotranspiration data, which are needed in the equations, are calculated from air temperature; in the software either by Thornthwaite [14] or Blaney-Criddle [15] method, in the paper by Thornthwaite only.

3 RUNOFF PROPHET SOFTWARE

Based on the above described methods, an application software Runoff Prophet was developed. Runoff Prophet is a rainfall-runoff model for simulations of river flow in catchment confluence. According to [16], it can be classified as a deterministic conceptual lumped model with monthly computing time step. The calculation requires no geographical characteristics of the catchment but its area. Firstly, the model is calibrated using historical climate and hydrological data, then it is possible to simulate future discharge based on predicted climate data.

Before hydrological modelling in a confluence of chosen river catchment, it is necessary to obtain required input data. Datasets related to the basin need to include monthly mean air temperature, monthly total precipitation and monthly mean water flow for calibration; monthly mean air temperature and monthly precipitation for subsequent prediction.

For the chosen catchment, localization is done first, which means specifying the basin area and geographical location of meteorological stations, where the used climate data comes from. Then a method of evapotranspiration estimation is chosen and settings of differential evolution parameters and upper limits of calibration coefficients are made. Afterwards, calibration can be run. Its success rate is quantified by the value of the NS criterion. When the model is calibrated for the basin, predictions can be executed.

4 APPLICATION

To evaluate the suitability of Runoff Prophet for different natural conditions, five categories of river basins covering the basic range of their typology were determined: i) agricultural (A), ii) forest (F), iii) mountain (M), iv) urbanised (U), and v) combined land use (C). For each category, several experimental catchments with hydrological measurements in its confluence were selected.

As area of interest, the Morava River basin was selected, being the Czech part of the first order basin of the Danube River. There are over 150 hydrometric stations in the Morava Basin owned by the Czech Hydrometeorological Institute (CHMI). Based on CHMI expert hydrologist's professional knowledge of the territory, 17 catchments of various size were chosen and classified into pre-determined categories (Fig. 1).

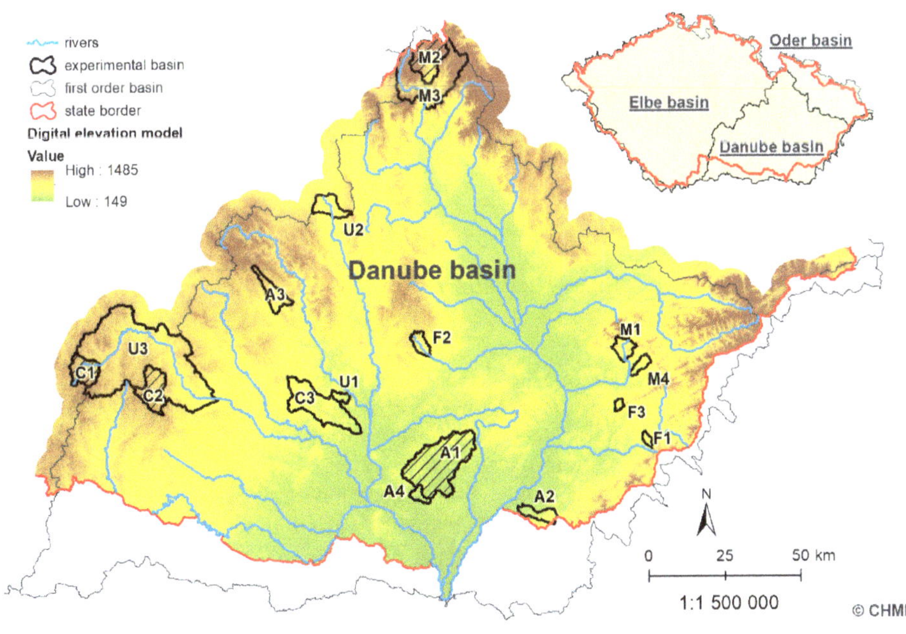

Figure 1: Chosen experimental catchments (CHMI).

For the purposes of final assessment, each basin was assigned an ID according to its category. Rudimentary catchment characteristics quantified were: total area in km^2, mean slope and proportions of forest, built-up area and arable land including orchards (Table 1).

Table 1: Selected catchments.

Catchment characteristics						Catchment area proportion (CORINE Land Cover 2018)		
Basin	Basin ID	Hydrometric station	River	Total area (km^2)	Mean slope (%) (DMR 5G)	Forest (%)	Built-up area (%)	Arable land + Orchard (%)
Agricultural	A1	Bořetice	Trkmanka	285.81	7.6	23.4	5.1	62.3
	A2	Petrov	Radějovka	41.02	7.1	39.6	1.8	26.4
	A3	Rožná	Nedvědička	57.19	5.6	19.7	4.4	58.9
	A4	Velké Pavlovice	Trkmanka	304.59	7.6	22.4	5.1	62.0
Combin	C1	Batelov	Jihlava	73.48	4.7	29.0	4.3	32.9
	C2	Brtnice	Brtnička	95.32	4.4	37.8	3.4	46.8
	C3	Želešice	Bobrava	181.52	7.5	48.6	9.0	38.0
Forest	F1	Bojkovice nad nádrží	Kolelač	9.76	8.0	52.1	0.0	13.3
	F2	Opatovice nad nádrží	Malá Haná	30.92	7.7	60.2	2.4	31.5
	F3	Řetechov-Pradlisko	Ludkovický potok	8.45	12.2	47.4	5.2	6.3
Mountain	M1	Kašava nad nádrží	Dřevnice	36.70	14.3	75.9	2.5	0
	M2	Habartice	Krupá	109.35	13.6	61.4	0.8	0
	M3	Raškov	Morava	349.79	13.9	51.7	1.1	0.2
	M4	Slušovice	Všemínka	21.22	13.3	53.4	9.4	0.4
Urbanised	U1	Bystrc	Vrbovec	15.02	9.3	56.3	17.7	21.4
	U2	Hradec nad Svitavou	Svitava	55.20	3.7	18.3	11.1	61.1
	U3	Ptáčov	Jihlava	962.71	5.5	31.0	5.2	42.4

4.1 Data processing

Hydrological and climate data for model calibration and validation had to be obtained for each catchment. CHMI provided observed data from the selected hydrometric stations and from all the current weather stations located in the area of chosen river basins and their surroundings. The climatological stations within the station network are not evenly spatially distributed and in addition, the size of the catchments varied, therefore the total number of usable weather stations for the individual basins was different. For each catchment, time series from 1 to 5 stations with temperature measurement and from 2 to 10 stations with rainfall measurement were identified and processed in order to find the best combination of temperature and precipitation input data for the hydrological model.

Hydrological data from 17 hydrometric stations corresponding to the chosen catchments were required in the form of mean monthly river flow (m^3/s) and were exported from the CHMI Hydrology department database. Datasets of mean monthly air temperature (°C) from 32 stations and total monthly precipitation (mm) from 63 stations were processed by CHMI Meteorology and climatology department. For the purposes of the experiments, different combinations of stations with different length of calibration/validation datasets had to be tested and evaluated. One combination contained data from one gauging station, one temperature station and one rainfall station. The combinations were created separately for each basin according to following criteria.

Mean water flow time series had to have at least 13 years of gauging data with simultaneous measurement of both temperature and rainfall data at one or more related weather stations. Temperature time series had to be at least 13 years long (during the period with available discharge data). If more stations fulfilled this condition, the ones with at least the same time series length as river flow were chosen. If there were more such stations, the one which was not evaluated as distant from the basin was chosen. If more stations with temperature data fulfilled the criteria, the combinations were prepared from all of them. Rainfall datasets had to be at least 13 years long (with simultaneous water flow measurement). Each satisfactory temperature dataset was combined with each satisfactory precipitation dataset and was complemented by the water flow data.

After applying all the criteria, the following number of weather stations was chosen (Table 2). Two of the stream gauging stations (A1 and C2) did not have measurement records long enough, hence they were excluded from further testing.

Table 2: Number of chosen discharge, temperature and precipitation weather stations.

ID	Q Chosen	TEMP Possible/Chosen	RAIN Possible/Chosen
A1	0	-/-	-/-
A2	1	1/1	3/3
A3	1	1/1	3/3
A4	1	3/1	9/7
C1	1	3/1	2/1
C2	0	-/-	-/-
C3	1	3/2	3/3
F1	1	5/1	7/6
F2	1	3/2	5/4
F3	1	2/1	2/1
M1	1	5/2	8/7
M2	1	5/1	10/9
M3	1	5/1	10/9
M4	1	5/2	8/8
U1	1	2/1	4/3
U2	1	1/1	3/3
U3	1	3/1	4/2

Each combination had to be divided into different variants of calibration and validation dataset length. Minimal length of calibration time series was set to 10 years, further variants gradually being increased by five years (10, 15, 20, etc.). Calibration to validation dataset ratio was 3:1 as tested in [8], validation series length was rounded to the nearest whole year (3, 5, 7, etc.). Validation time series preceded the relevant calibration with no time gap (Fig. 2).

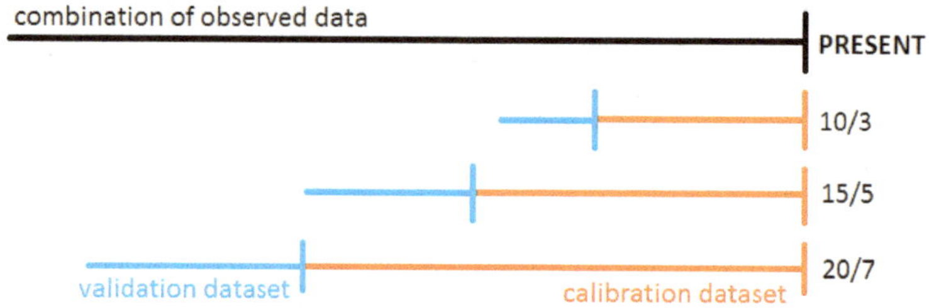

Figure 2: Scheme of calibration and validation datasets.

4.2 Runoff Prophet calibration and validation

For each basin, all the created combinations were evaluated using Runoff Prophet. Firstly, the calibration was run with calibration dataset as an input and its efficiency expressed by the NS was recorded. Then the validation was performed using temperature and rainfall data from the relevant validation dataset. Its NS was calculated from the comparison of predicted water flow and water flow from the validation dataset time series.

For all the experiments, Runoff Prophet settings were constant. Thornthwaite method for estimating evapotranspiration was used, the parameters of differential evolution and upper limits of calibration coefficients were both set to default.

5 RESULTS AND DISCUSSION

For the basins, all the datasets were evaluated by NS of calibration and NS of relevant validation. If NS of calibration and also validation exceeded 0.5, the combination was rated as satisfactory [17]. For each catchment, the results of all the combinations were listed according to the highest value of validation NS. Optimal data sources and lengths are specified in Table 3.

From the 15 considered basins, eight were evaluated as satisfactory. For most of them, 10 years of calibration (three of validation) time series were evaluated as the optimal length. In one case, 15 years (five for validation) was the most suitable option, in three cases, it was 25 (six for validation).

The catchments had a different size across the defined categories, therefore there was a reason for examining its impact on model performance (Fig. 3). Coherent low model efficiency was recognised on the smallest basins. Calibration/validation on three catchments with area of less than 15 km^2 was considered unsatisfactory. In two cases, NS of validation was very close to the satisfactory limit (0.45 on F3 and 0.48 on F1) and in all cases the basins fell into the problematic categories (F and U), as explained below.

A correlation between land use on the basin and satisfactory result of model calibration and validation was found across the categories. The catchments with more than 10% of urbanised area gave unsatisfactory result (two cases). Also the catchments with ratio of arable and orchard land of more than 60% had the resulting NS evaluated as unsatisfactory (two cases).

Table 3: Results of best validation/calibration in the representative basins.

Catchment characteristics			Optimal combination			NS (Best VAL)	
ID	Hydrometric station	Satisfac-tory	TEMP	RAIN	Length (years)	CAL	VAL
A2	Petrov	Y	Strážnice	Strážnice	10	0.83	0.73
A3	Rožná	Y	Bystřice nad Pernštejnem	Bystřice nad Pernštejnem	15	0.70	0.64
A4	Velké Pavlovice	N	Pohořelice	Ždánice	10	0.45	0.05
C1	Batelov	Y	Počátky	Třešť	25	0.66	0.69
C3	Želešice	Y	Troubsko	Moravské Bránice	10	0.69	0.68
F1	Bojkovice nad nádrží	N	Vizovice	Uherský Brod	25	0.63	0.48
F2	Opatovice nad nádrží	N	Protivanov	Blansko	10	0.66	0.40
F3	Řetechov-Pradlisko	N	Vizovice	Horní Lhota	10	0.73	0.45
M1	Kašava nad nádrží	Y	Vizovice	Zlín, Velíková	10	0.75	0.73
M2	Habartice	Y	Šerák	Malá Morava, Sklené	10	0.78	0.53
M3	Raškov	N	Šerák	Králíky	10	0.63	0.46
M4	Slušovice	Y	Vizovice	Zlín, Velíková	10	0.72	0.74
U1	Bystrc	N	Troubsko	Zastávka	25	0.33	-0.39
U2	Hradec nad Svitavou	N	Gajer, Janov	Gajer, Janov	10	0.55	0.38
U3	Ptáčov	Y	Hubenov	Hubenov	10	0.69	0.54

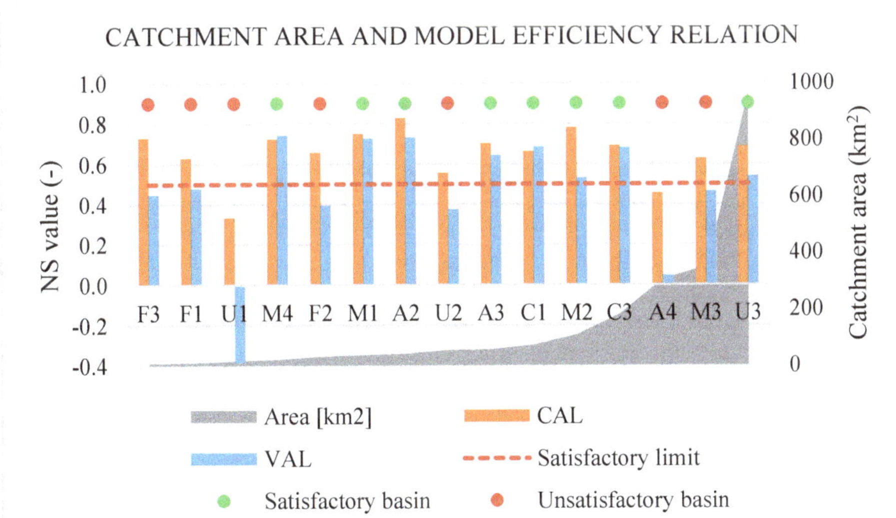

Figure 3: Relationship between catchment size and model efficiency (NS of calibration/validation from optimal combination).

The representative basin categories were evaluated separately and the results of NS were rated in more detail according to [17]:

5.1 Agricultural river basin (A)

Two out of three catchments were evaluated as satisfactory. A2 had very good NS of calibration and good NS of validation, A3 good NS of calibration and satisfactory result of validation. For A4, the model did not succeed neither in the calibration, nor validation.

5.2 Forest river basin (F)

Runoff Prophet did not satisfactory simulate any of the chosen forest catchments. Even when NS coefficients of calibration were good or satisfactory, the validations were not successful and were evaluated between 0.4 and 0.48.

5.3 Mountain river basin (M)

Three out of four catchments achieved satisfactory results. In all three cases, model performance was very good or good in terms of calibration and good or satisfactory in terms of validation. The unsatisfactory basin M3 was very close to fulfilling the limits, achieving NS of 0.63 for calibration and 0.46 for validation.

5.4 Urbanised river basin (U)

One of the three catchments fulfilled the satisfactory limit. Interestingly, it was Ptáčov (U3), the largest basin with a total area of 962.71 km^2.

5.5 Combined land use river basin (C)

Both of the combined catchments met the satisfactory limit with a good value of NS for calibration as well as validation.

According to the NS results, Runoff Prophet performed very well on combined land use river basins and mountain river basins. Good results were also achieved for agricultural basins. Urbanised river basin category was problematic, with variable results and only one satisfactory case. The model was completely unsuccessful regarding forest basins, with consistent results of NS slightly below the specified limit.

Providing a clear explanation of poor results in the last two categories is not easy. Hydrological modelling on urbanised basins can be disputable due to anthropogenic influence on the rainfall-runoff process. Natural monthly relationships between precipitation, temperature and discharge can be disrupted by municipal water management, in particular water consumption, retention and draining. Regarding the forest basins, for assessment it is interesting to compare the catchments characteristics with very good resulting category of mountain basins. The ratio of forest land cover is ≥45% for both categories. Mountain basins have an even larger proportion of forest area. However, a significant difference between them is in the mean slope. Mountain basins have a mean slope of more than 13%, the forest basins are not so steep – mean slope in this case is less than 13%. There is also a difference between arable and orchard land area, where mountain category includes catchments with 0 to 1% of such area, whereas the chosen forest basins have 6–32% of arable land area. Large area of the forest can have significant impact on rainfall-runoff process and the insufficient coverage of weather stations may lead to inaccurate description of the precipitation nature. The model confirmed the presumption that in the mountain basins, precipitation is transformed into discharge faster due to steeper slopes and the significance of forest microclimate and small water cycle declines.

Another question is whether the number of available weather stations for the basin increases the accuracy of model calibration/validation results. Fig. 4 shows the number of satisfactory (+) and unsatisfactory (-) basins for different number of stations. A relationship was then searched for between the possible stations, i.e. before applying selection criteria, and also chosen stations.

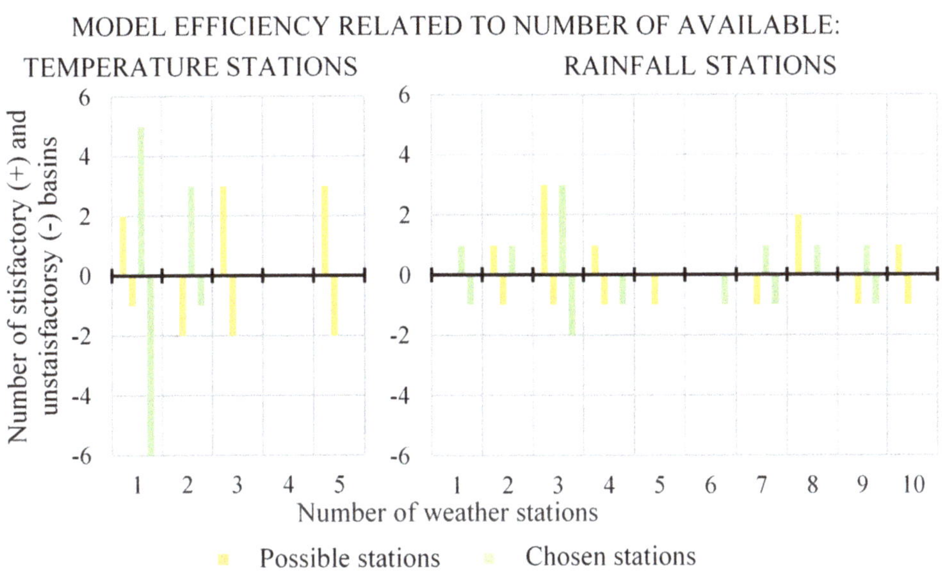

Figure 4: Model efficiency in relation to the number of available weather stations.

Concerning the temperature data, there is no visible correlation between the number of possible stations in the basin and the model result. For the chosen stations, more positive ratio of basins was evaluated as satisfactory for two stations (3:1) than for one (5:6). It was important to find out, if a good result of a basin is truly a consequence of more variants of used temperature data or if it is caused by catchment characteristics and easier conditions for hydrological modelling. On four catchments, two temperature stations passed the defined criteria and were used for creating variant calibration/validation datasets. Table 4 shows the results of best validation NS for two variants of temperature data source. As can be seen in the table below, satisfactoriness of the basin is same for both temperatures in case of all events. There is also no big difference between resulting NS of calibration and validation, ranging from 0–0.03 for calibration and 0–0.06 for validation.

With regards to the rainfall data, the relationship between the number of possible and chosen stations and satisfactoriness of the basin is non-determinable due to a high variability in the number of stations (Fig. 4). For more detailed insight, the relationship between available stations number and the resulting calibration and validation NS was examined. As can be seen in Fig. 5, there is no clear correlation between the number of chosen rainfall stations and the resulting NS or basin satisfactoriness.

Table 4: NS results for basins with two variants of chosen temperature stations.

Basin ID	TEMP 1			TEMP 2			Difference of NS (Best VAL)	
	NS (Best VAL)		Satisfactory	NS (Best VAL)		Satisfactory		
	CAL	VAL		CAL	VAL		CAL	VAL
F2	0.66	0.40	N	0.69	0.33	N	0.03	-0.06
C3	0.69	0.68	Y	0.71	0.61	Y	0.02	-0.07
M1	0.75	0.73	Y	0.75	0.71	Y	0.00	-0.02
M4	0.72	0.74	Y	0.73	0.74	Y	0.00	0.00

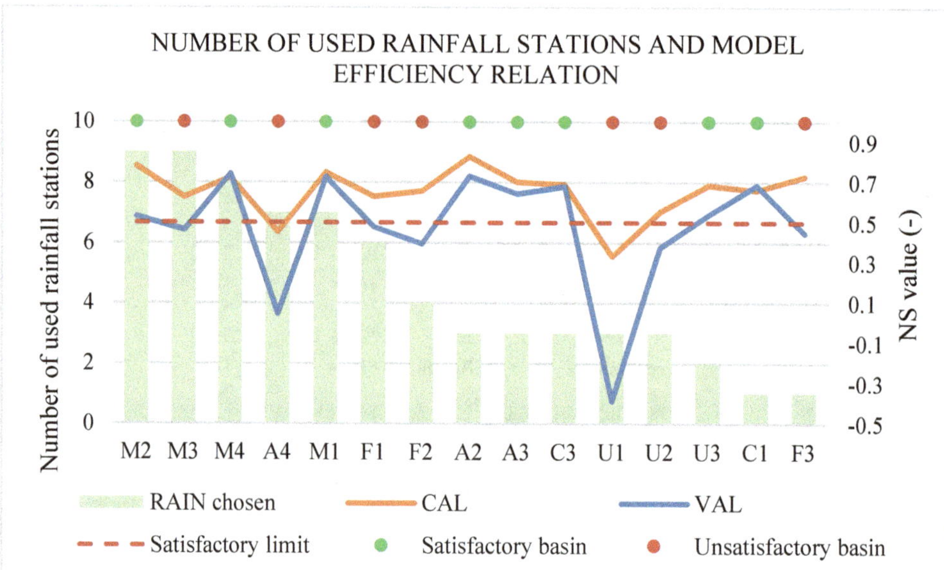

Figure 5: Relationship between the number of chosen rainfall stations and model efficiency.

It is important to note that the calibration for the experimental basins was run with the same default settings. It is likely that different basins would achieve reliable results with different upper limits of calibration coefficients and parameters of differential evolution.

Also a more robust method of model efficiency verification would be appropriate with more variants of calibration time periods and more validation datasets for each variant.

6 CONCLUSION

Runoff Prophet, a novel rainfall-runoff model developed at Brno University of Technology, was systematically tested to confirm its wide-range usability in various hydrological conditions in the Czech Republic. A set of representative catchments from the Morava River Basin, Czech part of the Danube Basin, was chosen. For these 15 catchments, Runoff Prophet was calibrated and validated across different combinations of input data source and time series length. Based on the achieved value of Nash–Sutcliffe model efficiency coefficient, an optimal combination of air temperature and precipitation input data was determined. Model performance for each catchment was rated as satisfactory or unsatisfactory. The results were evaluated from the perspective of river basin category, land use characteristics, area and number of available weather stations.

The results of the experiments presented in this paper showed that Runoff Prophet is more suitable for hydrological modelling of mountain river basins, agricultural river basins and river basins with combined land use. Its application on forest and urbanised river basins did not lead to satisfactory results. It was also shown, that there was no provable relationship between the size of catchment area and model calibration success. Based on the number of available stations, no effect of quantity of both air temperature and rainfall considered data sources on model performance was found. From the total number of 15 experimental basins, 8 were modelled with satisfactory results and 4 were very close to the satisfactory limit. In the light of catchments variability, this can be declared better-than-expected outcome, which could be further enhanced by using more robust methods of optimal data combination testing.

For further studies, presented relationship between catchment land use and model efficiency should be taken as only indicative. For varying basins, individual testing of model performance should be made and evaluated. Ability of conceptual hydrological model to reflect rainfall-runoff process in the catchment strongly depends on the availability of high-quality input data. Continuous hydrological gauging in the river catchment confluence is the basis for such input. Air temperature is relatively spatially consistent, therefore the number of stations for Morava Basin was adequate even when strict time series length requirements were applied. In contrast, precipitation is much more variable in space, thus the density of station network was not sufficient (about one station for every 100 km^2 in the Morava Basin), in particular for smaller catchments. The success of calibration on individual catchment does not rely on the number of usable rainfall data sources, but on the availability of weather stations with sufficiently long data series suitably located in the area of interest. The suitability is not determined only by geographical proximity of the station, but especially by its appropriate location for representing the nature of rainfall for the area of the basin as a whole.

Software testing presented in this paper produced good results and confirmed, that Runoff Prophet is an effective tool for conceptual hydrological modelling. It also brought about many suggestions for further improvement of the methods used and upgrades of the software.

It was proved, that Runoff Prophet and the original water balance equations are usable in variable hydrological conditions of the Czech Republic. It is expected, that the software is suitable for rainfall-runoff modelling of diverse basins with similar conditions and characteristics as presented in the experiments described in this paper, whether in areas with temperate climate or worldwide.

ACKNOWLEDGEMENTS

We would like to express our appreciation to colleagues from the Czech Hydrometeorological Institute: G. Knozova for operative supply of climate data, P. Coufal for help with catchments classification, B. Barta for providing catchments characteristics and map and P. Benacek for help with data processing.

This research was funded by Climate Change Uncertainty Propagation in Hydrologic and Water Management Applications research project number (FAST-S-18-5341). It is also the result of CHMI Department of Hydrology project of the internal intention.

Runoff Prophet can be downloaded and used for non-commercial purposes as freeware from following website: uvhk.fce.vutbr.cz/cs/software.php.

REFERENCES

[1] Klimatická změna, CzechGlobe (Ústav výzkumu globální změny AV ČR v.v.i.), Brno. www.klimatickazmena.cz. Accessed on: 4 Dec. 2018.

[2] Czech Hydrometeorological Institute, Tisková zpráva 22.11.2018, Brno. portal.chmi.cz/informace-pro-vas/tiskove-zpravy/2018. Accessed on: 4 Dec. 2018.

[3] HEC-HMS. www.hec.usace.army.mil/software/hec-hms/.

[4] NOAA National Weather Service, NWSRFS User Manual Documentation, Silver Spring. www.nws.noaa.gov/ohd/hrl/nwsrfs/users_manual/htm/xrfsdocpdf.php. Accessed on: 11 Jan. 2019.

[5] HySoft Brno. www.hysoft.cz.

[6] SWAT. https://swat.tamu.edu/.

[7] Beven, K., Lamb, R., Quinn, P., Romanowicz, R. & Freer, J., TOPMODEL. *Computer Models of Watershed Hydrology*, ed. V.P. Singh, Water Resources Publications: Colorado, pp. 627–668, 1995.

[8] Knoppová, K., Srážko-odtokový proces v podmínkách klimatické změny. Master's thesis, Brno University of Technology, 2018.

[9] Wang, G.Q. et al., Simulating the impact of climate change on runoff in a typical river catchment of the Loess Plateau, China. *Journal of Hydrometeorology*, **14**, pp. 1553–1561, 2013.

[10] Knoppová, K. & Marton, D., Srážko-odtokový model pro dlouhodobé prognózy v podmínkách klimatické změny. *Hospodaření s vodou v krajině*, 2018.

[11] Knoppová, K., Runoff Prophet, Nástroj pro dlouhodobé hydrologické prognózy v podmínkách změny klimatu. *Konferencie mladých odborníkov*, 2018.

[12] Marton, D. & Knoppová, K., Robust reliability assessment of water reservoir under uncertainty of climate change. *HIC 2018 – 13th International Conference on Hydroinformatics*, pp. 1316–1323, 2018.

[13] Zelinka, I., *Umělá inteligence v problémech globální optimalizace.* BEN-technická literatura: Prague, 2002.

[14] Appendix 1: Thornthwaite Method: Water balance on the Central Rift Valley; Universitat politecnica de Catalunya, UPCommons: Portal de coneixement obert de la UPC, Barcelona. https://upcommons.upc.edu/bitstream/handle/2117/89152/Appendix10.pdf?sequence=3&isAllowed=y. Accessed on: 15 Jan. 2019.

[15] Chapter 3.1.3 Blaney-Criddle Method: Chapter 3: Crop water needs; Food and Agriculture Organisation of the United Nations, Rome. www.fao.org/docrep/S2022E/s2022e07.htm#. Accessed on: 15 Jan. 2019.

[16] Abbott, M.B. & Refsgaard, J.C., Deterministic models. *Distributed Hydrological Modelling*, Kluwer Academic Publishers: Dordrecht, pp. 27–30, 1996.

[17] Moriasi, D.N., Arnold, J.G., Van Liew, M.W., Bingner, R.L., Harmel, R.D. & Veith, T.L., Model evaluation guidelines for systematic quantification of accuracy in watershed simulations. *Transactions of the ASABE*, **50**(3), pp. 885–900, 2007.

ADEQUACY OF GEOTEXTILE TUBE DEWATERING IN THREE RIVER REMEDIATION SCENARIOS

GERBEN VAN DEN BERG[1] & FILINTO ALEXANDRE PINTO DE OLIVEIRA[2]
[1]TenCate Geosynthetics Netherlands B.V., The Netherlands
[2]Geosin, LDA, Portugal

ABSTRACT

Natural processes and human interference can lead to undesirable sedimentation of a magnitude difficult to deal with in an economical, ecological and societally acceptable manner. Showcasing various examples throughout the world, criteria to choose an adequate river remediation technology will be discussed, in three distinct scenarios: (1) the case of natural processes leading to a build-up of contaminated sediments; (2) the case of industrial activity leading to contamination; and finally (3) the case of disaster management. For the first time we will discuss the approach taken in Portugal to clean the Tago river at Vila Velha do Rodão where a layer of sediment with high cellulose levels covering the bottom on both sides of the Porta de Rodão at the heart of the Natura 2000 Site and UNESCO World Heritage listed natural park had to be removed and dewatered. Other cases of river remediation projects will be presented as well, such as the case of Fox river (USA) and Zutphen (the Netherlands), where contaminated sediments were used to rebuild the entrance of a river port. We will also touch upon the potential for tube dewatering technology as a response during disaster management and the hurdles encountered in the actions around two red mud disasters. The first dating from 2010 in Hungary at Ajka where a dam breach in a containment facility released 1.1 Mm^3 of red sludge into a widespread area, polluting agricultural land and water ways. The second at the Rio Doce at Vale Samarco in Brazil in 2015, where dewatering bags were used to build a dam using the red mud spilled into the river.

Keywords: tube dewatering, contaminated sediments, remediation, TenCate Geotube®, sediment use.

1 INTRODUCTION

Although sediment load management in rivers can reduce erosion and sediment settlement at inappropriate places, maintenance dredging remains an indispensable tool in keeping the desired functionality of a river intact. When sediment mixes with industrial or municipal waste streams, the approach taken increases in complexity, cost and often duration. Questions like problem ownership and project funding assume a political edge due to the effects of contaminants on human health, flora and fauna. Decision-taking processes take longer with the increased need to investigate the gravity of the problem, the best way to remediate, and the effects of the remediation itself on public health, flora and fauna, as well on the public opinion. With increasing awareness of the need for a clean environment and to streamline these processes, legislation, rules and best-practice guidelines have been developed in the past years by waterbodies, dredging associations, and environmental policy makers on regional, national and international levels.

Now that dumping at sea of contaminated sludges is forbidden in Europe, alternative solutions are even more needed especially in deltas and near ports where most river sediments end up. Landfill potential is limited and expensive, so that alternatives for dumping are actively studied and practiced (Gregoire [1]). Although primary construction material is scarce in many areas, dredged sediment is considered to be waste and to fall under the waste directive of the EU. For reasons of scale, durable "local" treatment alternatives are easier to envisage close to a port than along a river since treatment can be organised economically and on a larger scale in the vicinity of the port, such as the facilities created in Antwerp, Hamburg and close to Rotterdam. River-sediment deposits may be wider spread and are often more

WIT Transactions on Ecology and the Environment, Vol 234, © 2019 WIT Press
www.witpress.com, ISSN 1743-3541 (on-line)
doi:10.2495/RBM190161

unpredictable in volume. Regulations surrounding land and upland dewatering sites are very restrictive, particularly for land that is near streams, rivers and lakes. For the same reasons, finding appropriate re-use possibilities seems even more difficult.

Taking a holistic approach and on the basis of several examples of realized projects at different places in the world, this paper discusses decision criteria for the use of geotextile-tube dewatering technology in three distinct scenarios: the case of natural processes leading to a build-up of contaminated sediments, the case of industrial activity leading to contamination and finally the case of disaster management.

2 GEOTEXTILE-TUBE DEWATERING

Tube dewatering technology helps in the de-silting waterways in a very effective way. Simultaneous to the dredging operation, the dredged volumes are massively reduced by separating liquids from solids and contaminants. In the final stage the consolidated contaminated solids are either transported off-site or are left in place and capped such that an in-situ landfill is created. Depending on the degree of contamination the land can then be re-used for appropriate purposes. On land, but also semi-submerged, tubes have been used for land reclamation, the creation of sedimentation areas, shore-line refurbishment and current diversion.

Tube dewatering technology consists of various successive phases in volumes that are determined mainly by the capacity of the dredge. Sludge is pumped up and introduced directly into large, permeable geosynthetic containers in which solid-liquid separation takes place. Very often polyelectrolytes are used to accelerate the separation and to fully use of the permeability of the geotextile fabric. Some fabrics achieve a permeability of up to $18m^3/m^2/min$. From the moment pumping starts, the water drains through the pores of the containers while the particles are retained. The proper addition of polyelectrolyte and the correct choice of fabric lead to high particle-retention rates of up to 99% of even the finest particles below 63 microns. By adsorption, undissolved contaminants like heavy metals will at the same time be retained effectively. After the fill stage, the geotextile containers are left to drain for an optimum result. For sandy sludges this may take a day or so, for silts and clays it can be a matter of weeks. The containers can then be removed, at a time convenient to the contractor, or they can be used as an integral part of a structure of a geotechnical nature, like the foundation of a parking lot or a park lawn.

3 TO USE OR NOT TO USE, THAT IS THE QUESTION

To determine if such a tube application is possible or not, criteria are defined on a case-to-case basis. Apart from the obvious budgetary constraints and legislative boundaries, geotechnical and technical questions need to be answered satisfactorily. These technical questions concern the quantity of material available, the characteristics of the material, the type of equipment needed, and the time and space limitations that are given to realize the project with sediment. In most cases potential sediment use will be weighed against other options. So a strong case needs to be made.

Prior to the use of sediments in a construction it should be clear what the purpose of the construction will be, and whether the contamination will contribute to, or limit, the functionality and the life expectancy of that construction. The demands on the construction and the materials used should be defined, as well as the way it can be built and with which selection of materials.

Potential sediment uses are dependent on the sediment characteristics (mean diameter, organic matter content, shear strength, density, Atterberg limits, etc.) and linked to the ecological and toxicological acceptability locally as well as the economic benefits that can be found close to its source for products or features made using the sediments. Beneficial use of dredged material based on scientific knowledge can be more practical, cost effective and environmentally advantageous when integrated with watershed-planning processes and regional strategies for managing sediment resources (Oliveira [2]).

In many applications where sediments are used, such as for agricultural topsoil, land creation, bricks and pavements, dikes and berms, contamination, storage, handling and low shear strength are issues to be tackled. To overcome low shear strength, cement or other fill material may be incorporated in the dredged slurry for stabilisation purposes. Apart from the local availability and handling of large amounts of cement, stabilisation alone does not address the other issues, particularly storage of slurry prior to use. The ability of geotextile tubes to immediately dewater, store, contain and support dredged materials can address these problems for a variety of applications [...] (Howard et al. [3]).

Before dredging starts, it is necessary to know how much sludge is to be treated, how much dilution is necessary to make it pumpable, how long it will take to move the slurry and what is needed in terms of effort, time and space for dewatering to get to the stage of the intended use. Time and space are often the most basic requirements. Because filling and dewatering are continuous and occur simultaneously if the right fabric is chosen, the project run-time can normally be shortened using environmental suction dredging in combination with tube dewatering technology. The reduction of volume obtained has a direct effect on the space requirements. Less space is needed when less water is stored. To reach this goal, good dredging and pumping practice, sampling and monitoring during the dewatering as well as the right choice of equipment, like the right choice of fabric, polymer, shutters and pumps, help to avoid costly mistakes.

4 THE TUBE

The choice for a specific fabric and a specific tube configuration will determine to a large extent the effective water release and particle retention. The latter is an important criterion since contaminant retention depends to a large extent on the retention of particles below 60 μm. In the absence of tests for large-scale sludge filtration, standard geotextile test methods are used, the results of which do not always reflect the real situation. There is for instance no test for clogging. Thus, despite equal (clear) water permeability according to EN ISO 11058 and a comparable O90 opening size, a tape-tape fabric will still clog faster in a dredging environment than a tape-monofilament fabric that achieves similar test results. To reduce the tensile forces on a plugged tube, several tape-tape fabric suppliers advise releasing the water by opening an unused fill port. It goes without saying that particle retention, especially of the finest (potentially contaminant-holding) particles is than no longer guaranteed, so that recontamination can take place. For this reason tape-tape fabric should be excluded on contaminated remediation sites.

Another misused geotextile criterion is tensile strength. Seams being the weakest part of a tube, seam strength, as an indicator of the overall strength of a tube, is more important than the tensile strength of the fabric itself.

Proper dimensioning of the bags, based on tests performed prior to the works, helps in getting the maximum amount of water out while effectively retaining fine particles. It also contributes to the continuity of the dredging operation and shortens the project run time, just as the correct choice of fabric, the correct flocculant dosing and the most appropriate layout will also do. Under-dimensioned tubes will potentially lead to turbulence inside the bag, to flocs breaking down and the fabric clogging, with poor dewatering as a result. Over-dimensioning on the other hand may potentially lead to uneven filling of the tube, excessively slow filling, and to failure to raise the tube to the maximum fill height. As demonstrated by

Stephens [4] correct but simple testing of the right fabric, accelerant and sludge combination before the design stage can help to prevent these problems and enables proper dimensioning of the hydraulics, the bags and the site.

5 THE SITE

Indeed the dimensions of a site can be determined on the basis of the quantity of sludge that is to be treated, the amount and characteristics of the dry solids in that sludge, and the available project run-time. The exact dimensions of a site can then be reduced by taking into account the hourly and daily capacity of the dredge, the capacity of the flocculant dosing unit, the required consolidation time per fill cycle, and the stacking potential of the consolidated tubes. If stacking is possible the dewatering can take place on a much smaller footprint.

For stacking, the tube content needs to have sufficient bearing capacity, as should the subsoil on which the tubes are laid out. Lawson [5] identified six external and four internal stability modes; local instability, geotextile bag rupture, erosion of fill through the geotextile bag and deformation of the contained fill. Zhu [6] divided the local instability into sliding, bearing failure, foundation settlement and slope movement. Fig. 1 shows these four geotechnical failure states potentially endangering a dewatering platform with an additional particular form of local instability that results from the fact that tubes are not sufficiently close to one another; V-notch deformation. This causes over-tensioning of the fabric of the upper tube when it slides into an oversized gap between two tubes below. Irregular consolidation, which can be caused by poor dewatering of the fabric, or varying settlement characteristics and bearing capacities of different fill materials in the same bag, has a similar effect to erosion of fill through the geotextile bag, but leaves an uneven base which can cause great stability problems for subsequent layers.

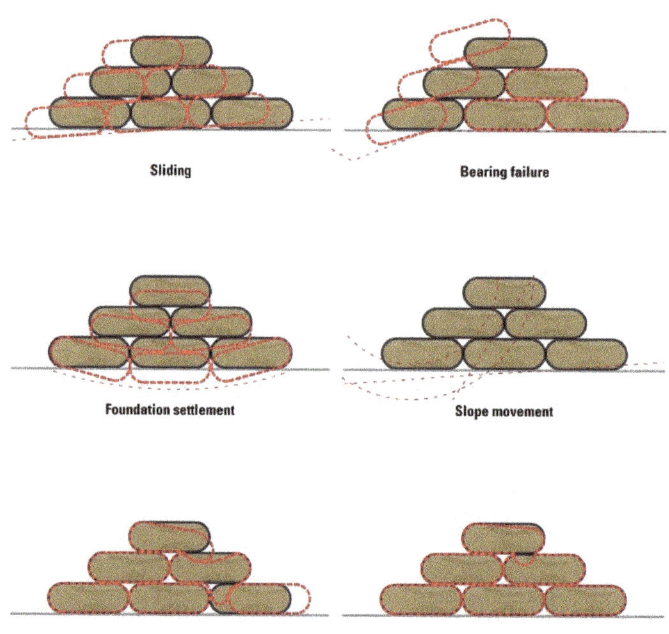

Figure 1: Failure states to avoid on dewatering platforms.

Once the sludge and sediment characteristics are known, pumping cycles determined, tubes dimensioned and space requirements set, construction and clearing of the dewatering site should also be taken into consideration in the design stages. Access to the tubes, stability of roads, and transport distances to potential landfills influence the total project run time, but also have financial consequences that can influence the decision for or against re-use.

Now that we have discussed general technical and geotechnical criteria, we will show in the case studies how other criteria also play a role and interact with each other. The fact for instance that tube dewatering technology provides dewatering and consolidation simultaneously, reduces cost. Further cost reduction is obtained through reduced installation cost, reduced material transport, carbon footprint, equipment mobilisation, operation and dismantling cost. These benefits increase when the tubes enable use of the sediments.

6 ZUTPHEN

Zutphen is a clear case of erosion of a riverside caused by wave action created by ships entering the port. The solids sedimented just meters from the shore line.

The city of Zutphen in The Netherlands is located on River IJssel, a tributary of River Rhine. The old industrial harbour De Mars was refurbished as part of a restoration plan for the entire industrial area. The refurbished port would have to be able to accommodate ships with a draught of 2.8 meters, but the actual situation was that the silting of the entrance of the port led to smaller ships running aground in front of the harbour at low tide. Dredging the entrance to the original depth was clearly required. At the same time the riverbank needed to be restored to prevent silting in a few years after the termination of the works.

Figure 2: The entrance of the port with TenCate Geotube® units installed.

Bathymetrical studies and sediment analyses showed that a total of 6,000 m^3 of Class 2 sediment and 12,000 m^3 of Class 3 contaminated sediment had to be dredged. The Class 3 contaminated sediment would have required placement in confined disposal facilities.

For the restoration of the riverbanks, new building materials would have to be brought in to raise the level of the embankments and to protect against waves at high water levels. The TenCate Geotube® dewatering solution solved both problems in an economical and environmentally friendly manner.

Stretching out along the 400 m riverbank that had to be restored, four sheet-pile compartments each 100 m wide were erected. The sheet piles delimited the dewatering platform from the River Ijssel. Not only they would give protection in case of rising water levels, but each adjacent basin could potentially serve as a temporary storage area in which the effluent water from the tubes could first be checked before being discharged into the river. In fact the opening in the sheet-pile construction through which the effluent water ran back freely into the river never needed to be closed, since at no point during the works did the quality of the effluent exceed the limits defined by the Ministry of Transport, Public Work and Water Management.

Each compartment held 4 tubes sized in relation to the sludge dewatering mass balance and the total surface area. With a volume of 400 m^3/h at a 10% solids concentration, the environmental dredge filled two tubes at a time. Due to the accurate in-line dosing of chemical polyelectrolytes, the dry solids content of the material in the tube reached 65% within four days. After all the material had consolidated and dredging was finished the tubes were covered with a soil backfill.

The local authorities and the ministry can look back on a successful project that cost them only 70% of the cost using conventional construction methods. The harbour and its entrance were successfully dredged and the riverbank was restored with the use of local sediments.

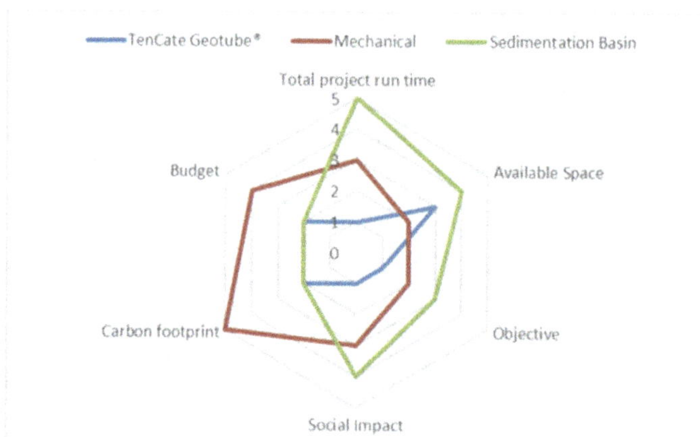

Figure 3: Chart comparing geotextile tube dewatering, mechanical dewatering and sedimentation basins on 6 criteria in relationship to the Zutphen project.

Re-use of contaminated sludge as a construction material has become inevitable. The Dutch Decree Soil Quality pragmatically integrates all aspects from dredging to application of soils and sludge, and favours local decision making based on risk assessment whereby the quality of the sludge mixture is regarded in relation to the function of the area it will be applied in and the development of the area as a whole. The potential risks of local standards are calculated on the basis of the soil function, along with an estimate of the degree of exposure (high, low average or contact with the soil or crop consumption of the soil) and the desired degree of protection of people, ecology and agriculture. The decree provides the generic norms, but local authorities can deviate from those norms as long as they stick within the stipulated framework. The decree contributes to a more positive way in which sludge is perceived; one no longer talks about slightly polluted soil, but sludge suitable for the class A "residential area" function.

7 FOX RIVER

In the state of Michigan, USA, it was primarily the major PCB-contamination by several paper mills which required intervention at the head of the lower Fox River in Northeast Wisconsin. PCB is a known carcinogen in animals and a probable carcinogen in humans. After many years of study, the funds became available, turning what had been up to then a political and financial issue into a technical issue. When tube dewatering tests showed that 99.7% to 99.9% of solids could be captured and the contaminants were retained inside the monofilament-tape fabric bag, tube dewatering was selected as the most economical process for the first part of the works.

To be able to extend the consolidation period to 45 days on average, stacking proved crucial. Here a combination technique was used with clarifiers in front of the tubes, to evacuate immediately most of the free water as well as debris larger than 1/8", so that hoses would block and reduce the flow towards the tubes. This set up helped to reduce space even more than was already obtained by stacking the geotextile bags that were 18.3 m in circumference and up to 50 m long. To avoid overdosing, polymer dosing was based upon the bed density inside the thickener unit.

Table 1: Hanging bag test results showing the capture rate of suspended solids in time.

Hanging Bag Test Total Solids	**Logon No. 2**	**Removal**	**East Ditch**	**Removal**
Influent total aolids	6.50%		14.70%	
1 min filtrate suspended solids	12.5 mg L	99.80%	178 mg L	98.80%
10 min filtrate suspended solids	14.0 mg L	99.78%	18 mg L	99.87%
1 hour filtrate suspended solids	5 mg L	99.92%	4.4 mg L	99.97%
Total Solids (after 10 days	24.20%		35.30%	
Hanging Bag Test Total Solids	**West Ditch**	**Removal**	**West Ditch No polymer**	**Removal**
Influent total aolids	12.10%		12.10%	
1 min filtrate suspended solids	2752 mg L	77.20%	174 mg L	99.56%
10 min filtrate suspended solids	36 mg L	99.70%	0.9 mg L	99.99%
1 hour filtrate suspended solids	11 mg L	99.91%		
Total Solids (after 10 days	50.40%		52.80%	

Table 2: Test results showing contaminant concentrations with almost undetectable levels of PCBs traces in the effluent.

Constituent		Concentration range of sludge	Concentration in TenCate Geotube® Effluent
PAH compounds	ug/kg	1.500 - 49.000	Not detected (1-5 ug/L)
PCB's (total)	ug/kg	100.000 - 140.000	0.14 ug/L
PCB's (total (with polymer)	ug/kg	100.000 - 140.000	Not detected (0.1 ug/L)
Cadmium	mg/kg	2.6 - 7.9	Not detected (0.5 ug/L)
Lead	mg/kg	512 -495	9.4 ug/L
Copper	mg/kg	326 -343	Not detected (10 ug/L)
Arsenic	mg/kg	7.3 - 7.7	5.5 ug/L
Mercury	mg/kg	3.8 - 18.4	0.14 ug/L
Zinc	mg/kg	3.700 - 3.700	13 ug/L

8 TAGUS RIVER (RIO TEJO)

Another example of a successful remediation project where geotextile tube dewatering provided the solution is found at Vila Velha do Rodão. The River Tagus suffered from eutrophication, a problem related to large amounts of cellulose and the presence of paper industry in this part of Portugal. A layer of sediment with excessively high organic levels covered the bottom on both sides of the Portas de Rodão site at the heart of the Natura 2000 Site and UNESCO World Heritage listed as natural area.

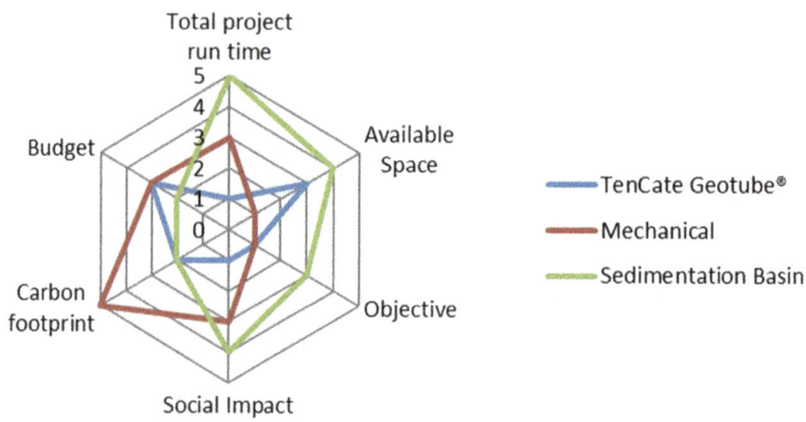

Figure 4: Chart comparing geotextile tube dewatering, mechanical dewatering and sedimentation basins on 6 criteria in relationship to the Portas de Rodão project.

Several dewatering treatment solutions were studied and tube dewatering technology finally came out best. Other solutions such as decantation basins required a lot of space and time. Also for reasons of impracticalities at the operational level as well as on social impact, decantation basins were discarded. At the other end of the spectrum, mechanical technologies were not preferred either. To mechanically treat such a large amount of sludge on an hourly basis would require powerful machines, or several of these. In addition it would require the construction of a monitoring basin for quality assurance. And finally, from a budget standpoint, either the cost had to go up in order to keep the total project run time reasonable, or the project would last longer than the tube dewatering solution for a similar budget. Geotextile tube dewatering was chosen over other methods for its rapidity, unlimited capacity of flow reception, adaptability and low carbon footprint.

As part of the solution to remediate and improve the ecosystem, the cellulose-holding sediments were hoovered in with a suction dredge operated by an experienced team of Portuguese divers. The pump used in this project was equipped with a sensor that allowed the height between the hose and the bottom of the river to be regulated. Several water jets disaggregated the sludge and kept the particles within the suction circle area. This was the most precise way of making sure that only sediment was removed while sand and rock remained in place. At the same time it allowed the slurry to be diluted sufficiently to achieve good flocculation later on in the process. All dredging equipment was mounted on a floating platform and monitored by professional divers. Once taken in, the slurry was pumped over 1.2 km to a temporary dewatering platform built by a consortium of highly specialized Portuguese companies.

This platform consisted of 30,000 m^2 open area adjacent to the river and still within the national park. This made the whole project rather sensitive and called for a state-of-the-art and safe dewatering platform. Because of this and the temporary nature of the facility, two separate 2.500 m^2 basins were created on the site surrounded by drainage ditches. To avoid any contaminant leakage, the entire site was lined with a sandwich of a puncture-resistant nonwoven, a liner and a drainage layer. All materials came from the same supplier, which simplified their procurement and installation. For the same reason, the special membrane arrived on site fully welded, which also offered additional assurance on the quality of the joints.

A reception tank, initially required by the owner as a means of control over the flocculation, did in practice not add to the quality and even slowed down the operation, since it had to be cleaned out twice. From this tank, standing on a hill some 4 meters over the tubes, on average 225 m^3/hr of flocculated sludge were gravity-fed into several of the eight connected tubes (0.7 bar).

Figure 5: Influent, flocked influent, effluent.

As discussed, contaminant retention depends on the retention of particles below 60 μm. To achieve this, the choice of fabric as well as good flocculation is crucial. At the River Tagus the tube was made of a tape-monofilament fabric. The polymer was specially selected, prepared and dosed from a mobile unit sufficient to treat 12 m^3/hr of solution. Dosage and concentration could be adapted to changes in the dry solids volume by touch screen. Due to the excellent water release and particle retention the effluent could be discharged directly into the river.

Figure 6: Removal of the dewatered bag content.

Each TenCate Geotube® unit had a capacity of 367 m³. Over 90,000 m3 of slurry at 1,8% dry solids was treated this way. In just 1.5 months the goal of 40% average dry solids was reached and the tubes were removed, leaving the spot clear as if nothing had happened.

"Operação Tejo 2018" was the first public environmental remediation project in Portugal using tube dewatering. The extensive experience of contractors with this technology in other fields contributed largely to its success, as was recognized by the local community and national environmental authorities.

9 AJKA AND RIO DOCE

That local experience is key, is all the more true when it comes to rapid response management as we will briefly illustrate discussing the aftermath of the disasters in Ajka, Hungary in 2010 and the Bento Rodrigues dam disaster in Brazil five years later. In both cases an earth dam gave way, liberating millions of cubic meters of iron sludge from metal production. In both cases tube dewatering and consolidation technology could have been part of the solution. Red sludge can be dewatered large scale in tubes, and tubes can be turned into the cores of dams to redirect the flow, protect an area or to create new storage areas. In both cases a direct line was established at the highest level. In Hungary the absence of an experienced local partner proved to be critical and locally practiced solutions were pursued. In Brazil tube technology was used. Several designs were made to build a containment dam with tubes creating a sludge disposal area. Apart from issues like ease of access and communication, accurate mobilisation of means for testing to prove that a proposed solution will actually work, was crucial. Without proper testing, the effective use of tube technology is merely based on the experience of its installers. Experienced contractors know however, that without testing, the effectiveness of tube dewatering cannot be guaranteed. Thus experience should be one of the considerations when choosing for any technology, but certainly when opting for tube technology.

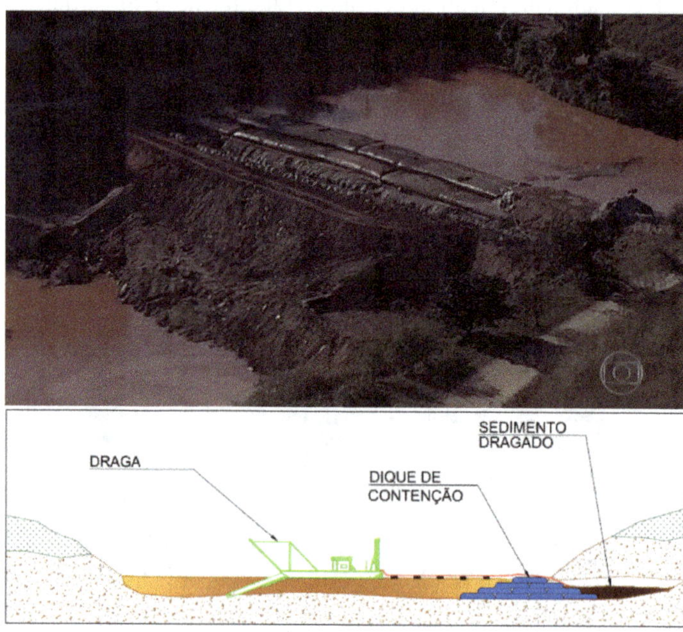

Figure 7: Picture and drawing of the tube dam.

10 CONCLUSION

When evaluating the different possibilities to dewater sludge at least six main elements can be weighed in relationship to each other:

- objective;
- budget;
- carbon footprint;
- available space;
- social impact;
- total project run time.

When considering the use of sediments for building a structure, one should also look at:

- elementary questions of technical and geotechnical nature;
- the advantages and benefits of alternative or regular ways to build the construction;
- its purpose and the stability it should have;
- the right flocculant formula and dosage;
- the choice of the appropriate fabric;
- the most adequate dimensioning of the tube;
- good dredging practice;
- the interaction between the former three bullet points;
- local presence of the need for the solution;
- local presence of qualified installers.

We have demonstrated looking at the various cases how tube dewatering and containment can bring an answer to the question what to do with large quantities of contaminated sediments in river basins. In two of the cases present, Zutphen and Brazil, there was a need for a geotechnical structure locally and a possibility to use dredged sediments for that purpose. In such cases high social acceptability, low carbon footprint, savings made in transportation and by the reduction of the project run time, tube dewatering and consolidation is a valid choice.

ACKNOWLEDGEMENTS

I wish to acknowledge the help provided by Mr Tom Stephens and Mr Howard Murray.

REFERENCES

[1] Gregoire, P., Valoriser 300,000 m^3 de sédiments de dragage en France? C'est possible! *INTERSOL, International Conference on Soils Sediment and Water*, Lille, 2014.

[2] Oliveira, E., Beneficial use of dredged material in agricultural land. *9th International SedNet Conference,* Krakow, p. 54, 2015.

[3] Howard, I.L., Vahedifard, F., Williams, J. & Timpson, C., Geotextile tubes and beneficial reuse of dredged soil: applications near ports and harbors. *Proceedings of the Institution of Civil Engineers – Ground Improvement*, **171**, pp. 1–32, 2017. DOI: 10.1680/jgrim.17.00059.

[4] Stephens, T., Canal do Fundão contaminated sediments GDT analysis versus actual full scale project result. *Geotechnical Special Edition 211 Geo-Frontiers 2011*, Dallas, pp. 2131–2140, 2011.

[5] Lawson, C., Geotextile containment for hydraulic and environmental engineering. *J. Geosynthetics International*, **15**(6), pp. 384–427, 2008.

[6] Zhu, M., Geotextile tube dewatering. *Geostrata*, pp. 26–32, 2015.

CALIBRATION AND VALIDATION OF ET_0 THROUGH AN R-CRAN CODE IN AGRICULTURAL LANDS OF SOUTH-EAST SPAIN

ANTONIO JODAR-ABELLAN[1], MARCOS RUIZ-ÁLVAREZ[2] & JAVIER VALDES-ABELLAN[1,3]
[1]University Institute of Water and Environmental Sciences, University of Alicante, Spain
[2]Euro-Mediterranean Water Institute, University of Murcia, Spain
[3]Department of Civil Engineering, University of Alicante, Spain

ABSTRACT

Currently, water demands from urban and agricultural use are increasing, especially in arid and semi-arid regions, such as the Mediterranean. This situation is expected to become worse with the climate change projections for the region, increasing the pressure, in both quantity and quality, on fresh water resources. Evapotranspiration (ET_0) is a hydrologic variable with high uncertainty and considered incorrect in water balance estimations. However, its accurate assessment is essential to obtain the real value of available water to satisfy water demands, especially in extended agricultural areas such as the south-east of Spain. ET_0 can be obtained using different equations with different levels of input data requirements, among them the Penman–Monteith option is the one recommended by the FAO (PMFAO), but its input data requirements are high. On the other hand, there are simpler options, such as the Hargreaves equation ($ET_{0,HG}$), but there is not such a big agreement about its accuracy in the scientific literature. The main objection to the use of PMFAO is the lack of some of the required meteorological variables in most climate stations, forcing the use of simpler alternatives. This paper presents an R-CRAN code where the $ET_{0,HG}$, parameterized by Samani, is calibrated and validated with the Allen model considering 18 statistical contrasts. Both $ET_{0,HG}$ results (pre- and post-calibrated) are compared with daily, monthly and annual results of the PMFAO. All meteorological data was provided by the CA52 Cartagena La Aljorra weather station, managed by the Agricultural Information System of the Murcia region (SE Spain). The main results show that daily, monthly and annual $ET_{0,HG}$ results after the Allen calibration and validation are similar to the PMFAO. However, a moderate underestimation of $ET_{0,HG}$ compared to PMFAO was identified. To sum up, the presented R-CRAN code provides an alternative to apply the $ET_{0,HG}$ method with few meteorological input requirements and, once calibrated, can be applied to extended data networks in other regions.
Keywords: evapotranspiration, Hargreaves and Penman–Monteith FAO equations, Allen calibration, agricultural areas, R-CRAN code.

1 INTRODUCTION

Evapotranspiration is one of the most important processes in the water cycle; its knowledge is very important in the disciplines of hydrology, meteorology, agriculture and to carry out studies on impacts of climate change. The accurate estimation of ET_0 is especially relevant in semiarid areas where water shortages are the major obstacle to agricultural production, economic welfare and sustainable development (Hargreaves and Allen [1], Samani [2], Allen et al. [3], [5], Monteith [4]).

There are different definitions of this process, among which is the reference evapotranspiration (ET_0), which is defined as "the evapotranspiration rate of a reference surface", which occurs without water restrictions. The reference surface is a hypothetical crop with an assumed crop height (0.12 m), a fixed canopy resistance (70 s m^{-1}) and Albedo (0.23) (Allen et al. [3], Martínez-Pérez et al. [6]).

Nowadays, the Penman–Monteith FAO (PMFAO), achieved with grass crops, is the recommended method to estimate ET_0, which is considered as the reference potential evapotranspiration. However, the meteorological data required by this method are not

www.witpress.com, ISSN 1743-3541 (on-line)
doi:10.2495/RBM190171

available in most weather stations. In these cases, Allen et al. [7] recommended the estimation of ET_0 using the Hargreaves equation (Hargreaves and Allen [1]), which only requires maximum and minimum temperature data. The equation of Hargreaves is defined as:

$$ET_{0,HG} = C \cdot RA \cdot (Tmax - Tmin)^{0.5} \cdot (Tmed + 17.8), \quad (1)$$

where $ET_{0,HG}$ is the reference evapotranspiration (mm day^{-1}) estimated by Hargreaves; C is an empirical coefficient whose value is 0.0023; RA is the extraterrestrial radiation (mm day^{-1}) measured as ET_0 equivalent (mm day^{-1}); $Tmax$, $Tmin$ and $Tmed$ are the daily maximum, minimum and average temperature (°C). See Allen et al. [7] for details of its calculation.

The Hargreaves equation frequently overestimates ET_0 in humid regions and underestimates it in dry regions. Moreover, Hargreaves equation tends to overestimate ET_0 in low ET areas and to underestimates it in high ET areas (Droogers and Allen [8], Xu and Singh [9]). For these problems, it is necessary to carry out a regional calibration of the Hargreaves equation. In Allen et al. [3], [10] it was proposed a methodology to realize this calibration, which consists in a simple linear regression between the series of ET_0 obtained from the Penman–Monteith and Hargreaves methods, passing the regression line by the origin of coordinates and using the slope of the regression line obtained as an adjustment parameter between these two expressions:

$$ET_{0,PM} = b_0 + b_1 \cdot ET_{0,HG}, \quad (2)$$

where $ET_{0,PM}$ is the reference evapotranspiration (mm day^{-1}) estimated by Penman–Monteith equation; and b_0 and b_1 are, respectively, the y-intercept and the line slope of the above mentioned regression line.

Thus, the main purpose of this study consists on accomplish an extend R-CRAN code [11] to resolve the ET_0 Hargreaves expression (Hargreaves and Allen [1]), parameterized by Samani [2], and assess its accuracy with previously obtained series of the PMFAO-ET_0. In order to improve this accuracy, results from the Hargreaves equation are calibrated and validated with the Allen et al. [3] model considering several statistical contrasts.

2 STUDY AREA

The present study was developed with registers from the CA52 Cartagena La Aljorra weather station, located in the Cartagena municipality (province of Murcia). These areas belong to the Segura river basin district (SE Spain) which presents semiarid climate with irregular and scarce rainfalls (average precipitations of 374 mm/year, showing clear differences from NW mountain areas, around 1,000 mm/year, to coastal areas with less than 300 mm/year), high annual average temperatures (18–10°C), and a large number of hours of sun that generate high potential evapotranspiration rates (average ETP of 993 mm/year) and real evapotranspiration rates (average ETR of 335 mm/year). Detail information can be found in CHS [12], Gomariz-Castillo et al. [13] and Ruíz-Álvarez et al. [14]. Due to the mentioned climate features, irrigated agriculture developed highly in these areas especially since the 1970s (alongside the start of the Tajo-Segura interbasin transfer) and, nowadays, constitutes great water demands together with high water necessities required from the industrial zones and the extend and densely populated urban areas identified in the south-eastern Spain. In particular, the Segura river basin district showed, in 2015, a total water demand of 1762.1 hm^3/year, which 86.2% corresponded to agricultural water demands, 10.8% to urban demands, 1.7% to environmental demands, 0.7% to industrial demands do not dependent of urban supply networks, and finally 0.6% to irrigate golf courses (CHS [12], Albaladejo-García et al. [15], Grindlay-Moreno and Lizárraga-Mollinedo [16]). As a result of the recognized instability among scarce water inputs and great water demands, currently the Segura river basin district presents the highest water resources shortage of Europe (Custodio et al. [17], Melgarejo-Moreno et al. [18]), with several problems of aquifers overexploitation (227 hm^3/year estimated in 2015 by CHS [12]), political issues related with interbasin transfers from other watersheds (Molina and Melgarejo [19]), great rates of desalinated water and reused wastewaters (from desalination: 158 hm^3/year to supply agricultural and urban uses; from reused wastewaters: around 78 hm^3/year), etc. all of them as measurements to supply the large identified water demands (CHS [12], Melgarejo-Moreno and Molina-Giménez [20]). Therefore, the correct quantification of water losses, as in this case the evapotranspiration rates, result essential specially in these types of water-stressed areas.

Fig. 1 depicts the analysed weather station, CA52 Cartagena La Aljorra during the period 1 September 1999–31 December 2015, in the context of the SE Spain. This station belongs to the Agricultural Information System of the Murcia Region (in Spanish SIAM [21]), a sub-network of the Spanish Agroclimatological Information System for Irrigation (in Spanish SIAR [22]).

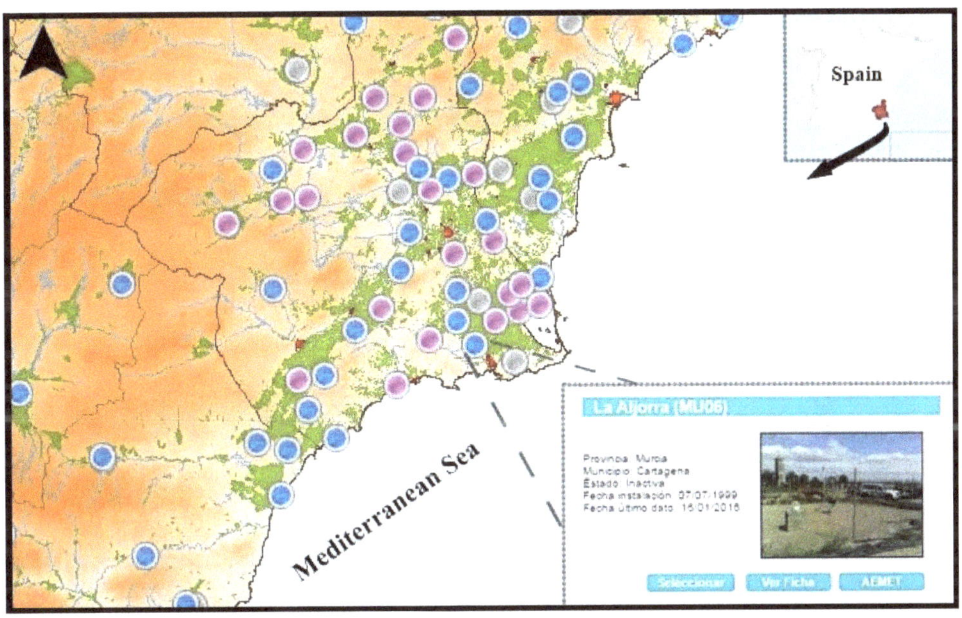

Figure 1: Study area showing the CA52 Cartagena La Aljorra weather station integrated in the SIAR network [22].

3 METHODOLOGY

In the following sections we present main items of the above mentioned R-CRAN code [11] to resolve the objectives of this study.

3.1 Previous steps

In the R-CRAN work routing, first of all, packages and libraries must be installed and loaded which depend of the study objectives. The "hydroGOF" library is necessary, among others, to perform next statistical contrasts (error estimations):

```
install.packages("hydroGOF","dep=T")
library(hydroGOF)
```

Next, daily information from CA52 Cartagena La Aljorra weather station (period 1 September 1999–31 December 2015) was imported in R-CRAN. This information contains data of Tmax, Tmin, Tmed, PMFAO-ET$_0$, etc:

```
dataET=read.table("data/dataorigin.csv",sep=";",dec=",",header=T,fil
l =T) # where fill=T is introduced by data gaps or NA.
```

Finally, imported daily data in R-CRAN is checked and processed before to start with the ETP equations:

```
names(dataET) # to display columns names.
dataET= dataET[,c(5,6,7,8,9)] #to conserve only columns of date, Tma
x, Tmin, Tmed, PMFAO-ET0.
names(dataET)=c("date","tmax", "tmed", "tmin","etp") # to change col
umns titles into appropriate titles (without spaces, etc.).
str(dataET) # to display the structure and type of information.
dataET$date=as.Date(dataET$date,format="%d/%m/%Y") # to convert the d
ate variable from factor or text type into date type. Expressions "%
d/%m/%Y" indicate day, month and year.
dataf2=data.frame(serie=seq(as.Date("1999-09-01"),as.Date("2015-12-3
1"),"days")) # to check the existence of all data. serie=seq to exte
nd serie from the initial date until the final date.
length(dataET$date)
## [1] 5930 # Thus, imported serie from SIAM present 5930 days with
data.
length(dataf2$serie)
## [1] 5966 #Thus, the period 01/09/1999-31/12/2015 present 5966 day
s. Therefore, imported serie from SIAM possess 36 days with data gap
s or NA.
dataf3=merge(dataf2, dataET,by.x="serie",by.y="date",all.x = T)#Thus
, we merge the matrix ''dataET'' and ''dataf2'' in a new matrix called ''
dataf3''. In the new matrix, days with data gaps are introduced with
a NA character.
length(dataf3$serie)
## [1] 5966 # to check the new matrix, called ''dataf3'', is correct.
subset(dataf3,is.na(dataf3$etp)) # to check the days with NA:
##          serie   tmax  tmed  tmin etp
## 204  2000-03-22   NA    NA    NA  NA
## 205  2000-03-23   NA    NA    NA  NA
## 206  2000-03-24   NA    NA    NA  NA
## 207  2000-03-25   NA    NA    NA  NA
## 208  2000-03-26   NA    NA    NA  NA
## 209  2000-03-27   NA    NA    NA  NA
## 210  2000-03-28   NA    NA    NA  NA
## 211  2000-03-29   NA    NA    NA  NA
## 212  2000-03-30   NA    NA    NA  NA
```

```
## 213  2000-03-31   NA    NA    NA  NA
## 214  2000-04-01   NA    NA    NA  NA
## 215  2000-04-02   NA    NA    NA  NA
## 216  2000-04-03   NA    NA    NA  NA
## 1403 2003-07-04   NA    NA    NA  NA
## (…)
```

3.2 Parameterized Hargreaves equation

In this section is shown the Hargreaves expression (Hargreaves and Allen [1]), parameterized by Samani [2], together with its parameters, partial and final equations:

```
ETHARG=function(latitude,date,tmed,tmax,tmin){
  #parameters: GSC=0.082
  #Julian day:
  djulian=as.numeric(julian(date,origin=as.Date(paste(substr(date,1,
4),"-01-01",sep=""))))
  #Partial equations:
  varpi=latitude*pi/180
  Dt=1+0.033*cos((2*pi/365)*djulian)
  delta=0.409*sin((2*pi/365)*djulian-1.39)
  omega=acos(-tan(varpi)*tan(delta))
  #Ra calculate:
  Ra=(24*60/pi)*GSC*Dt*(omega*sin(varpi)*sin(delta)+cos(varpi)*cos(d
elta)*sin(omega))
  Ra=Ra*0.408
  #ETHARG final equation:
  etharg=0.0023*(tmed+17.8)*(tmax-tmin)^0.5*Ra
  return(etharg)    }
source("FunctionHavers/FunHavers.R")
latdec=37+(40/60)+(35.9/3600) # Latitude is a variable necessary to
calculate the Hargreaves model. The CA52 Cartagena La Aljorra weathe
r station presents latitude of 37º 40´ 35.9´´ N, which was calculate
d from its UTM X and Y coordinates (SIAR [22]).
ETHARG(latitude=latdec,tmax=dataf3[1,"tmax"],tmed=dataf3[1,"tmed"],
tmin=dataf3[1,"tmin"],date=dataf3[1,"serie"])
## [1] 2.971072 # This is the Hargreaves ET0 result to the first lin
e of the ''dataf3'' data.frame.
for(i in 1:5966){
  dataf3$etharg[i]=ETHARG(latitude=latdec,tmax=dataf3[i,"tmax"],tmed
=dataf3[i,"tmed"],tmin=dataf3[i,"tmin"],date=dataf3[i,"serie"])
}# This is the Hargreaves ET0 expression to the complete ''dataf3'' da
ta.frame.
```

The complete daily serie of ET_0, obtained by the Hargreaves expression to the 5,966 days, underestimates data of PMFAO-ET_0 from the CA52 Cartagena La Aljorra weather station. Therefore, it is necessary to calibrate and validate the Hargreaves model.

3.3 Calibration and validation of the Hargreaves model

In this section the ET_0 Hargreaves expression (Hargreaves and Allen [1]), parameterized by Samani [2], is calibrated and validated with the Allen et al. [3] model considering 18 statistical contrasts:

```
modAllen=lm(dataf3$etp~dataf3$etharg) # This is the linear model of
Allen applied to the ''dataf3'' matrix in order to calibrate the ethar
g serie.
summary(modAllen) # The summary command displays important statistic
al data:
## Call:
## lm(formula = dataf3$etp ~ dataf3$etharg)
## Residuals:
##      Min       1Q     Median       3Q      Max
## -2.3380 -0.5335 -0.1595  0.3861  4.5037
## Coefficients:
##                            Estimate   Std.Error  t value Pr(>|t|)
## (Intercept)            0.32825     0.02197    14.94    <2e-16 ***
## dataf3$etharg     1.11502     0.00635   175.58    <2e-16 ***
## ---
## Signif. codes:  0 '***' 0.001 '**' 0.01 '*' 0.05 '.' 0.1 ' ' 1
## Residual standard error: 0.7717 on 5908 degrees of freedom
##    (56 observations deleted due to missingness)
## Multiple R-squared:  0.8392, Adjusted R-squared:  0.8392
## F-statistic: 3.083e+04 on 1 and 5908 DF,  p-value: < 2.2e-16
```

Regarding the presented coefficients, the y-intercept (b_0 in eqn (2)) shows a value of 0.32825. To remove this value the following expression is used:

```
modAllen=lm(dataf3$etp~0+dataf3$etharg)
summary(modAllen)
```

Likewise, the period with registers from the CA52 Cartagena La Aljorra weather station (1 September 1999–31 December 2015) is divided to validate and calibrate. The first third (1 September 1999–31 December 2004) is used in validation, and the others two thirds (1 January 2005–31 December 2015) in calibration. Therefore, both series (calibration and validation) are isolated by a subset:

```
modAllen=lm(subset(dataf3$etp,dataf3$serie>="2005-01-01")~0+subset(d
ataf3$etharg,
dataf3$serie>="2005-01-01"))
summary(modAllen)
```

In this subset the "Estimate" coefficient shows a value of 1.203716. This coefficient is the line slope (b_1 in eqn (2)) of the Allen model:

$$ET_{0,PM} = b_0 + b_1 \cdot ET_{0,HG}$$
$$ET_{0,PM} = 0 + 1.203716 \cdot ET_{0,HG}$$

Thus, in the "dataf3" data.frame a new column, correspond to the calibrated $ET_{0,HG}$, is added:

```
dataf3$ethargC = 1.203716*dataf3$etharg #The next picture depicts
the first 10 days of the analysed serie (5966 days):
```

	serie	tmax	tmed	tmin	etp	etharg	ethargC
1	1999-09-01	27.80	25.47	23.19	3.78	2.9710724	3.5763274
2	1999-09-02	29.92	26.07	22.15	5.60	3.8870677	4.6789256
3	1999-09-03	35.03	28.19	21.20	7.66	5.4031067	6.5038059
4	1999-09-04	28.95	25.65	20.73	4.68	3.9109005	4.7076135
5	1999-09-05	29.38	25.50	22.42	4.98	3.5635476	4.2894993
6	1999-09-06	29.84	25.83	22.28	5.15	3.7181941	4.4756498
7	1999-09-07	30.24	24.34	20.54	5.95	4.0412823	4.8645562
8	1999-09-08	27.30	24.03	20.79	6.18	3.2645775	3.9296242
9	1999-09-09	28.47	24.46	21.05	5.92	3.4974131	4.2098922
10	1999-09-10	28.00	24.25	21.08	5.85	3.3377863	4.0177468

Therefore, new values of the calibrated $ET_{0,HG}$ (ethargC in the previous picture) are more close to the PMFAO-ET_0 serie (etp in the picture) than the uncalibrated $ET_{0,HG}$ serie (etharg). However, a moderate underestimation of PMFAO-ET_0 is observed. Thus, to assess this adjustment, in calibration and validation, 18 statistical contrasts are considered.

3.3.1 Statistical contrasts

During calibration and validation, 18 statistical contrasts were used to assess error estimations between the calculated $ET_{0,HG}$ series and the reference serie (PMFAO-ET_0). These contrasts were: Mean error (ME), mean absolute error (MAE), mean squared error (MSE), root mean square error (RMSE), Normalized root mean square error (NRMSE in %), percent bias (PBIAS in %), RMSE-observations standard deviation ratio (RSR), standard deviation ratio (rSD), Nash–Sutcliffe efficiency (NSE), modified Nash–Sutcliffe efficiency (mNSE), relative Nash–Sutcliffe efficiency (rNSE), Willmott index of agreement (d), modified Willmott index of agreement (md), relative Willmott index of agreement (rd), Pearson's correlation coefficient (r), coefficient of determination (R^2), R^2 multiplied by the linear regression coefficient between simulations and observations (bR^2) and Kling–Gupta efficiency (KGE). Detail information about these statistical analysis and their numerical expressions can be found in Nash and Sutcliffe [23], Willmott [24], Gupta et al. [25], Krause et al. [26], Moriasi et al. [27] and Gupta et al. [28]).

The following commands calculate the above mentioned statistical contrasts based on the setting between the PMFAO-ET_0 daily series and the $ET_{0,HG}$ daily series:

```
gof(dataf3$etharg,dataf3$etp) #uncalibrated setting (period
01/09/1999-31/12/2015)
gof(sim=subset(dataf3$etharg,dataf3$serie>="2005-01-01"),obs=subset
(dataf3$etp,dataf3$serie>="2005-01-01")) #uncalibrated setting.
Calibration period (01/01/2005-31/12/2015).
gof(sim=subset(dataf3$ethargC,dataf3$serie>="2005-01-01"),obs=subset
(dataf3$etp,dataf3$serie>="2005-01-01"))#calibrated setting.
Calibration period (01/01/2005-31/12/2015).
```

```
gof(sim=subset(dataf3$etharg,dataf3$serie<"2005-01-01"),obs=subset
(dataf3$etp,dataf3$serie<"2005-01-01"))#uncalibrated        setting.
Validation period (01/09/1999-31/12/2004).
gof(sim=subset(dataf3$ethargC,dataf3$serie<"2004-01-01"),obs=subset
(dataf3$etp,dataf3$serie<"2004-01-01"))#calibrated          setting.
Validation period (01/09/1999-31/12/2004).
```

4 RESULTS AND DISCUSSION

4.1 Uncalibrated results

In this section uncalibrated results of $ET_{0,HG}$ at daily, monthly and annual time step are shown during the study period (1 September 1999–31 December 2015, Fig. 2). The following command is used to plot these results:

```
ggof(sim=dataf3$etharg,obs=dataf3$etp,dates=dataf3$serie,ftype="dma"
,leg.cex=1.2,FUN=sum,gofs=c("MAE","RMSE", "PBIAS", "NSE","d","r", "R
2"),legend=c("ET0 Harg.","ET0 PMFAO"),main="ET0 PMFAOvsETHargreaves"
,xlab="",ylab="mm",col=c("red","blue"),cex=0.5,cal.ini="2005-01-01")
```

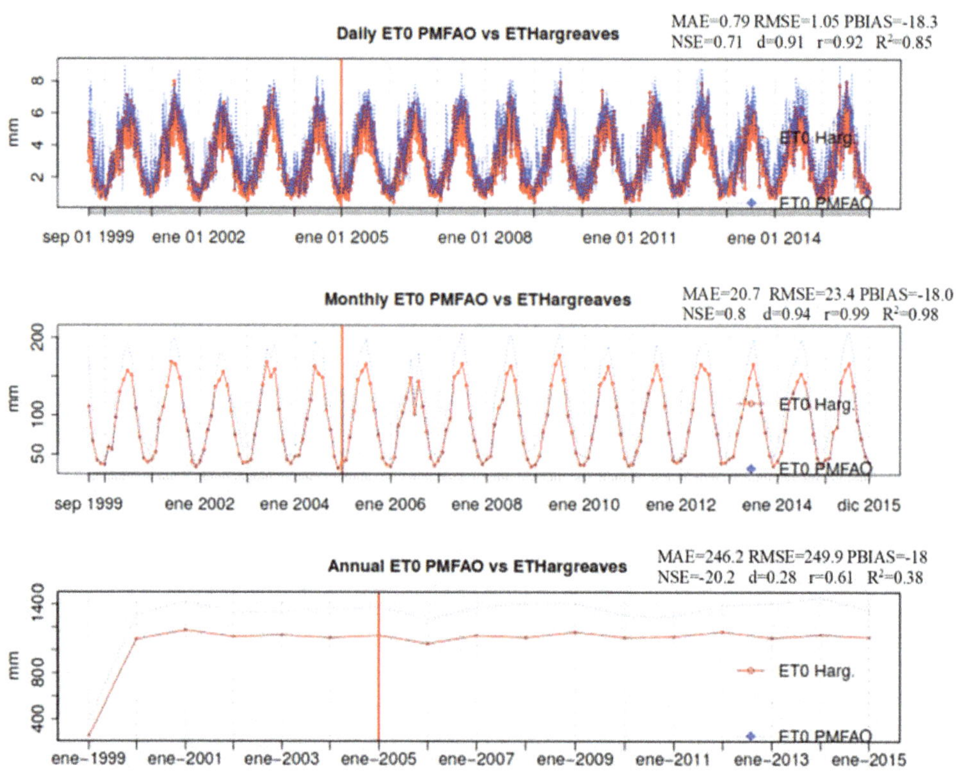

Figure 2: Uncalibrated results of $ET_{0,HG}$ Vs the reference serie (PMFAO-ET_0) at daily, monthly and annual time step during the study period (1 September 1999–31 December 2015).

As can be observed in Fig. 2, obtained series of $ET_{0,HG}$ depict a clear underestimation regarding the reference serie (PMFAO-ET_0). This circumstance can be verified trough the PBIAS statistic which results show values around -18% in the three time steps (daily, monthly and annual).

4.2 Calibrated results

In this section calibrated results of $ET_{0,HG}$ at daily, monthly and annual time step are shown during the study period (1 September 1999–31 December 2015, Fig. 3). The calibration and validation were carried out with the Allen et al. [3] model. The following command is used to plot these results:

```
ggof(sim=dataf3$ethargC,obs=dataf3$etp,dates=dataf3$serie,ftype="dma
",leg.cex=1.2,FUN=sum,gofs=c("MAE","RMSE", "PBIAS", "NSE","d","r", "
R2"),legend=c("ET0 Harg.Allen99","ET0 PMFAO"),main="ET0 PMFAO vs ETH
argreaves.Allen99",xlab="",ylab="mm",col=c("red","blue"),cex=0.5,cal
.ini="2005-01-01")
```

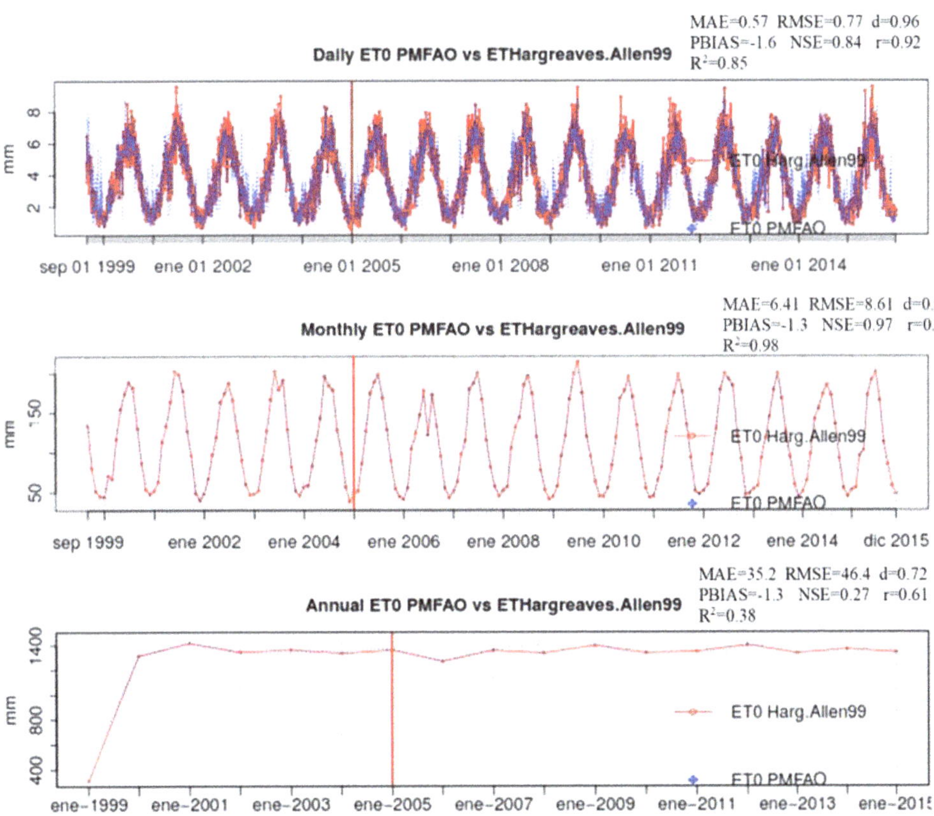

Figure 3: Calibrated results of $ET_{0,HG}$ Vs the reference serie (PMFAO-ET_0) at daily, monthly and annual time step during the study period (1 September 1999–31 December 2015).

In Fig. 3, calibrated series of $ET_{0,HG}$ depict closer to the PMFAO-ET_0 at the three time steps (daily, monthly and annual) than the uncalibrated $ET_{0,HG}$ series presented in Fig. 2. This circumstance can be verified through the statistic results closer to their optimal values (Moriasi et al. [27], Gupta et al. [28], Fig. 3). A good example is given by the percent bias (PBIAS) which show values between -1.3% and -1.6% (Fig. 3). Therefore, the PMFAO-ET_0 underestimation, previously identified in the uncalibrated $ET_{0,HG}$ series, has reduced considerably.

4.3 Statistic summary

This section depicts statistic results obtained during validation and calibration of the $ET_{0,HG}$ daily series and the contrast with their optimal values (Table 1).

Table 1: Results of the considered statistical contrasts during daily validation and calibration ($ET_{0,HG}$ Vs PMFAO-ET_0).

Statistical contrasts			Validation (1 September 1999–31 December 2004)		Calibration (1 January 2005–31 December 2015)	
Acronym	Range	Opt. value	Uncalibrated $ET_{0,HG}$	Calibrated $ET_{0,HG}$	Uncalibrated $ET_{0,HG}$	Calibrated $ET_{0,HG}$
ME	-∞...+∞	0	-0.66	-0.03	-0.69	-0.06
MAE	0...-∞	0	0.75	0.60	0.79	0.57
MSE	0...-∞	0	1.09	0.69	1.10	0.60
RMSE	0...-∞	0	1.04	0.83	1.05	0.77
NRMSE	-100%...100%	0	55.10	44.00	54.00	39.90
PBIAS	-100%...100%	0	-17.90	-0.90	-18.30	-1.60
RSR	0...+∞	0	0.55	0.44	0.54	0.40
rSD	0...+∞	0	0.83	1.01	0.82	0.98
NSE	-∞...1	1	0.70	0.81	0.71	0.84
mNSE	-∞...1	1	0.54	0.63	0.54	0.66
rNSE	-∞...1	1	0.74	0.73	0.75	0.74
d	0...1	1	0.91	0.95	0.91	0.96
md	0...1	1	0.76	0.82	0.75	0.83
rd	0...1	1	0.92	0.93	0.92	0.93
r	-1...1	-1&1	0.91	0.90	0.92	0.92
R^2	0...1	1	0.82	0.82	0.85	0.85
bR^2	0...1	1	0.67	0.80	0.68	0.82
KGE	0...1	1	0.74	0.90	0.73	0.92

As can be observed in Table 1, generally statistic contrasts achieved better results in the calibrated $ET_{0,HG}$ daily series, and even better in the calibrated $ET_{0,HG}$ to the period 1 January

2005–31 December 2015 (Table 1). It is considered that this finding is mainly due to the measure improvement often identified in meteorological stations during current periods.

5 CONCLUSIONS

Main conclusions of the present work are listed below:

- In the CA52 Cartagena La Aljorra weather station (SE Spain), the ET_0 Hargreaves series, parameterized by Samani [2] and simulated during the period with available data (1 September 1999–31 December 2015), showed initially a great underestimation of the Penman–Monteith FAO reference evapotranspiration at three time steps (daily, monthly and annual).
- After the $ET_{0,HG}$ validation (1 September 1999–31 December 2004) and calibration (1 January 2005–31 December 2015), with the Allen et al. [3] model, this variable enhanced its setting to the PMFAO-ET_0 reducing notoriously the above identified underestimation. Thus, results of the considered statistic contrast verified this improvement.
- Therefore, the presented R-CRAN code, to estimates the evapotranspiration variable, can be applied in extend areas over the world thanks to the high capability of R-CRAN to manage massive information (data).

ACKNOWLEDGEMENTS

This work was partially supported by the project GRE17-12 of the University of Alicante. Antonio Jodar-Abellan acknowledges financial support received from the Spanish FPU scholarship for the training of university teachers. In the same way, this work has been conducted within the Cátedra del Agua of the University of Alicante (https://catedradelaguaua.org/). Data provided by the SIAM and SIAR networks have been essential. Finally, the authors acknowledge the reviewers of the manuscript whose comments contributed greatly to improve this paper.

REFERENCES

[1] Hargreaves, G.H. & Allen, R.G., History and evaluation of Hargreaves evapotranspiration equation. *J. Irrig. Drain. Eng.-Asce.*, **129**(1), pp. 53–63, 2003.

[2] Samani, Z.A., Estimating solar radiation and evapotranspiration using minimum climatological data. *J. Irrig. Drain. Engi.*, **126**(4), pp. 265–267, 2000.

[3] Allen, R.G., Pereira, L.S., Raes, D. & Smith, M., Crop evapotranspiration: guidelines for computing crop water requirements. *FAO Irrigation and Drainage*, **56**, pp. 1–15, 1998.

[4] Monteith, J.L., *Evaporation and the Environment*, Cambridge University Press: Swansea, 1965.

[5] Allen, R.G., Jensen, M.E., Wright, J.L. & Burman, R.D., Operational estimates of evapotranspiration. *Agronomy Journal*, **81**, pp. 650–662, 1989.

[6] Martínez-Pérez, J.A., García-Galiano, S.G., Martin-Gorriz, B. & Baille, A., Satellite-based method for estimating the spatial distribution of crop evapotranspiration: sensitivity to the Priestley–Taylor coefficient. *Remote Sensing*, **9**(6), pp. 611–636, 2017.

[7] Allen, R.G., Pereira, L.S., Raes, D. & Smith, M., Evapotranspiración del cultivo. Guías para la determinación de los requerimientos de agua de los cultivo. *Estudio FAO Riego y Drenaje,* **56**, 2006.

[8] Droogers, P. & Allen, R.G., Estimating reference evapotranspiration under inaccurate data conditions. *Irrig. Drain. Syst.*, **16**, pp. 33–45, 2002.

[9] Xu, C.Y. & Singh, V.P., Cross comparison of empirical equations for calculating potential evapotranspiration with data from Switzerland. *Water Resour. Management*, **16**(3) pp. 197–219, 2002.

[10] Allen, R.G., Smith, M., Perrie, A. & Pereira, L.S., An update for the calculation of reference evapotranspiration. *ICID Bulletin*, **43**(2), pp. 35–92, 1994.

[11] R-CRAN code: A language and environment for statistical computing. https://cran.r-project.org/. Accessed on: 22 Nov. 2018.

[12] CHS, *Plan Hidrológico de la Demarcación del Segura 2015/21. Memoria.* Confederación Hidrográfica del Segura, Ministerio de Agricultura, Alimentación y Medio Ambiente, p. 806, 2015.

[13] Gomariz-Castillo, F., Alonso-Sarria, F. & Cabezas-Calvo-Rubio, F., Calibration and spatial modelling of daily ET_0 in semiarid areas using Hargreaves equation. *Earth. Sci. Inform.*, **11**(3), pp. 325–340, 2017.

[14] Ruíz-Álvarez, M., Gomariz-Castillo, F. & Alonso-Sarria, F., Análisis espacio-temporal de los cambios previstos en la evapotranspiración de referencia durante el siglo XXI en la Demarcación Hidrográfica del Segura (España) a partir de los modelos climáticos globales MPEH5 y MPEH5C. *Revista de Geografía Norte Grande*, **71**, pp. 35–58, 2018.

[15] Albaladejo-García, J.A., Martínez-Paz, J.M. & Colino, J., Evaluación financiera de la viabilidad del uso de agua desalada en la agricultura de invernadero del Campo de Níjar (Almería, España). *Información Técnica Económica Agraria*, **114**(4), pp. 398–414, 2018.

[16] Grindlay-Moreno, A.L. & Lizárraga-Mollinedo, C., Regadío y territorio en la Región de Murcia: Evolución y perspectivas de futuro. *Ciudad y Territorio*, **172**, pp. 281–298, 2012.

[17] Custodio, E. et al., Groundwater intensive use and mining in south-eastern peninsular Spain: Hydrogeological, economic and social aspects. *Science of the Total Environment*, **559**(1), pp. 302–316, 2016.

[18] Melgarejo-Moreno, J., López-Ortiz, M.I. & Fernández-Aracil, P., Water distribution management in South-East Spain: A guaranteed system in a context of scarce resources. *Science of the Total Environment*, **648**, pp. 1384–1393, 2019.

[19] Molina, A. & Melgarejo, J., Water policy in Spain: Seeking a balance between transfers, desalination and wastewater reuse. *International Journal of Water Resources Development*, **32**(5), pp. 781–798, 2015.

[20] Melgarejo-Moreno, J. & Molina-Giménez, A., *La Mancomunidad de los Canales del Taibilla en la Provincia de Alicante*, MCT e Instituto Universitario del Agua y de las Ciencias Ambientales: Alicante, pp. 1–279, 2017.

[21] SIAM, Agricultural Information System of the Murcia Region (Spain), http://siam.imida.es/apex/f?p=101:1000:3329167166355660. Accessed on: 1 Apr. 2018.

[22] SIAR, Spanish Agroclimatological Information System for Irrigation, http://eportal.mapama.gob.es/websiar/SeleccionParametrosMap.aspx?dst=1. Accessed on: 4 Apr. 2018.

[23] Nash, J.E. & Sutcliffe, J.V., River flow forecasting through conceptual models: Part 1. A discussion of principles. *J. Hydrology*, **10**(3), pp. 282–290, 1970.

[24] Willmott, C.J., On the validation of models. *Physical Geography,* **2**, pp. 184–194, 1981.

[25] Gupta, H.V., Sorooshian, S. & Yapo, P.O., Status of automatic calibration for hydrologic models: Comparison with multilevel expert calibration. *J. Hydrologic Eng.*, **4**(2), pp. 135–143, 1999.

[26] Krause, P., Boyle, D.P. & Base, F., Comparison of different efficiency criteria for hydrological model assessment. *Advances in Geosciences*, **5**, pp. 89–97, 2005.

[27] Moriasi, D.N., Arnold, J.G., Van Liew, M.W., Bingner, R.L., Harmel, R.D. & Veith, T.L., Model evaluation guidelines for systematic quantification of accuracy in watershed simulations. *Transactions of the ASABE,* **50**(3), pp. 885–900, 2007.

[28] Gupta, H.V., Kling, H., Yilmaz, K.K. & Martínez, G.F., Decomposition of the mean squared error and NSE performance criteria: Implications for improving hydrological modelling. *Journal of Hydrology*, **377**, pp. 80–91, 2009.

ASSESSING SOCIAL PERCEPTIONS OF EMERGING POLLUTANTS IN SOUTH-EAST SPAIN

MARIA DE LOS ANGELES BERNAL-ROMERO DEL HOMBRE BUENO[1], PABLO AZNAR-CRESPO[1,2], ANTONIO JODAR-ABELLAN[1], JOAQUÍN MELGAREJO MORENO[1] & DANIEL PRATS[1]
[1]University Institute of Water and Environmental Sciences, University of Alicante, Spain
[2]Department of Sociology, University of Alicante, Spain

ABSTRACT

EMPORE project, co-funded by LIFE Programme (LIFE15 ENV/ES/000598), aims to demonstrate an innovative, cost-efficient and highly replicable technology for the removal of Emerging Pollutants (EPs) from European Waste Water Treatment Plants (WWTPs). In action, C2, a set of indicators to monitor the socio-economic impact of EPs removal, was developed. Framed in this action, a robust assessment of the social perception of these contaminants was performed in the South-East Spain, where the use of reclaimed water is widespread due to the structural water scarcity. An expert consultation by means of a survey technique was held from November to December of 2018 with a sample size of 114 cases. The survey focused on regulation, risk perception, technologies and socio-economic impact. A descriptive analysis of 19 variables was carried out. In general, the results suggest that the presence of EPs can be a significant risk and destabilization factor in the reuse of wastewaters. The impact of new regulations concerning EPs awakes uncertainty among water experts. Aspects such as the price of water free of EPs, treatment costs or the potential of different technologies to remove these compounds are sources of ambiguity. Nevertheless, experts believe new regulations on EPs will positively influence public health, social confidence or environmental sustainability, among others. It highlights that experts agree that the future of the reuse of wastewater passes through the combination of technologies. These findings provide water planning policy makers with very useful information in the water resources management of the South-East Spain because recent European laws can set new limits of emerging pollutants in waters.

Keywords: emerging pollutants, social perception, descriptive statistical analysis, South-East Spain, water reuse.

1 INTRODUCTION

The implementation of the Water Framework Directive [1] raises a number of shared technical challenges for the Member States. The aim to reach good ecological status in every water body in Europe is far to be reached. Within this context, emerging contaminants (EPs) is one of the key topics to be addressed in the following years.

Directive [2] (article 26) defines EPs as "*pollutants currently not included in routine monitoring programmes at Union level but which could pose a significant risk requiring regulation, depending upon their potential ecotoxicological and toxicological effects and on their levels in the aquatic environment*". This term includes a wide variety of compounds, such as pharmaceutical products, flame retardants, personal care products and pesticides, among others.

The main objective of urban wastewater treatment plants (WWTPs) according to the Urban Waste Water Treatment Directive [3] and its extension Directive [4] is to remove suspended solids, organic matter, and, in sensitive areas, nutrients. However, many EPs are resistant to conventional treatments and are not removed neither altered in the water line and finally reach the aquatic media, concerning fact, especially when treated water is reuse.

Water reuse is an accepted practice in several European countries, especially in Spain, where the risks of water scarcity have enhanced the reuse of treated wastewater for agricultural, industrial and urban uses. Reuse of water for agricultural activities is also very

WIT Transactions on Ecology and the Environment, Vol 234, © 2019 WIT Press
www.witpress.com, ISSN 1743-3541 (on-line)
doi:10.2495/RBM190181

widespread in southern European countries, although it must also be highlighted that water reuse in these countries is also driven by tourism (BIO Deloitte [5]).

Water scarcity is especially accused in the South-East Spain, region where the reuse of treated water has been widespread to meet water needs for agriculture and tourism, especially in Murcia, Alicante and Almería. There are factors, such as quality, price and availability that may influence the social perception of the reuse of treated waters.

The project LIFE-EMPORE "Development of an efficient and sustainable methodology for EPs removal in WWTPs" (EMPORE-LIFE15 ENV/ES/000598) aims to demonstrate an innovative, cost-efficient and highly replicable technology for the removal of EPs from European WWTPs. The project is due to last 36 months, from September 2016 to August 2019, and its total budget is 1,783,824 €. The participants are: Laboratorios Tecnológicos de Levante S.L., IHE Delft, Entidad Pública de Saneamiento de Aguas Residuales de la Comunidad Valenciana (EPSAR), University of Alicante, Technology Institute on Metal-Processing, Wood, Furniture and related industries (AIDIMME), and CONSOMAR S.A.

The EMPORE project includes the design, construction and operation of a pilot plant that combines a set of technologies (conventional filtration, membranes filtration, advanced oxidation processes and electrochemical processes) to improve the quality of the secondary effluents of WWTPs regarding EPs. The demonstration is carried out in Benidorm (Alicante) and the developed technology will potentially be replicated and used throughout other places of the European Union with similar particularities (presence of EPs and water needs for reuse). Among other objectives, the project EMPORE aims to assess the socio-economic impact of the implementation of the demonstration plant in the local economy and also in other European regions with similar pollution problems. This article focuses on assessing the social perception of EPs in the South-East of Spain, region where the demonstration plant is operated.

2 STUDY AREA

The present study was developed during November and December of 2018. Most surveys were realised in the "XIV Technical Symposium on Water Treatment and Purification (*Jornadas* ESAMUR)" celebrated in the Lorca municipality (province of Murcia, Spain) from 21st to 22nd November 2018 (www.jornadasesamur.com/). This municipality belongs to the Segura river basin district (SE Spain) which presents semiarid climate with irregular and scarce rainfalls (average precipitations of 374 mm/year, showing clear differences from NW mountain areas, around 1000 mm/year, to coastal areas with less than 300 mm/year), high annual average temperatures (18–10°C), and a large number of hours of sun that generate high potential evapotranspiration rates (average ETP of 993 mm/year) and real evapotranspiration rates (average ETR of 335 mm/year). Detail information can be found in CHS [6], Gomariz-Castillo et al. [7] and Ruíz-Álvarez et al. [8].

Due to the mentioned climate features, irrigated agriculture developed highly in these areas specially since the seventy-eighty of the last century (alongside the start of the *Tajo-Segura* interbasin transfer) and, nowadays, constitutes great water demands together with high water necessities required from the industrial zones and the extend and densely populated urban areas identified in the south-eastern Spain. In particular, the Segura river basin district showed, in 2015, a total water demand of 1762.1 hm^3/year, which 86.2% corresponded to agricultural water demands, 10.8% to urban demands, 1.7% to environmental demands, 0.7% to industrial demands do not dependent of urban supply networks, and finally 0.6% to irrigate golf courses (CHS [6]; Albaladejo-García et al. [9]; Grindlay-Moreno and Lizárraga-Mollinedo [10]). As a result of the recognized instability among scarce water inputs and great water demands, currently the Segura river basin district presents the highest water resources shortage of Europe (Custodio et al. [11]; Melgarejo-Moreno et al. [12]), with several problems of aquifers overexploitation (227 hm^3/year estimated in 2015 by CHS [6], political issues related with interbasin transfers from other watersheds (Molina-Giménez and Melgarejo-Moreno [13]), great rates of desalinated water and reused wastewaters (from desalination: 158 hm^3/year to supply agricultural and urban uses; from reused wastewaters: around 78 hm^3/year), etc., all of them as measurements to supply the large identified water demands (CHS [6]). Therefore, assessing the social perception of emerging pollutants results especially relevant in areas, as the South-East Spain, where the use of reclaimed water is widespread due to the identified water scarcity. This particular situation has produced that, currently, the SE Spain depicts one of the highest desalinated and wastewater reused rates in the world in terms of production (m^3/day), alongside the developed technology and energy consumption (kW-h/m^3) as can be found in Zarzo and Prats [14] and Prats-Rico [15].

Fig. 1 depicts the Lorca municipality in the context of the Segura river basin district, together with the autonomous communities of the surveyed experts.

3 METHODOLOGY

An extensive review of the specialized literature was conducted (information not supply), consulting publications about water reuse, water resources management and different aspects of EPs (regulations, social perception, treatment technologies...). The information was useful to define the three basic aspects of this research: i) the conceptual gaps in EPs; ii) the scope and potential topics of the study; and iii) the profile of the experts to consult.

Figure 1: Study area showing the Lorca municipality in the context of the Segura river basin district (SE Spain) and the autonomous communities of the surveyed experts.

The information to assess the social perception of EPs was obtained through an expert consultation using the survey as a technique of data collection. The questionnaire consisted of three main thematic blocks: i) general aspects; ii) technological aspects; and iii) aspects of social, economic and environmental impact. Due to the scarce specialized literature on perception issues linked to EPs, a generalist approach was implemented to gather information on this topic.

The fieldwork consisted in the realization of face-to-face and individual structured interviews (survey). The survey took place in Lorca municipality from 21st to 22nd November 2018 during the "XIV Technical Symposium on Water Treatment and Purification (*Jornadas* ESAMUR)". This congress gathered more than 300 experts in the field of sanitation and wastewater treatment. Due to the geographical proximity, this congress recruited many experts from the provinces of Murcia and Valencia. These regions of the South-Eastern of Spain present the highest levels of water reuse and desalination of this country, and of the greatest of Europe, as a consequence of the identified water deficit (Custodio et al. [11]; Melgarejo-Moreno et al. [12]). Thus, although the statistical stratification of sample was not an objective, a significant number of cases from the regions most linked to water reuse and more potentially affected by EPs was managed. Altogether, 114 case studies were accomplished. This sample size was sufficient to satisfy the requirements of the nonparametric study. Once the primary data was obtained, the processing of data was carried out through the statistical software SPSS [16]. Fig. 2 represents the general methodological diagram achieved in this study.

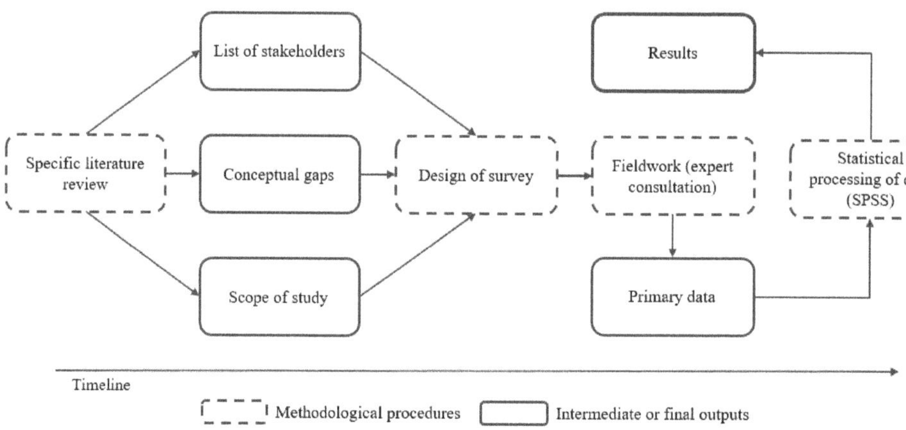

Figure 2: Methodological process of the present study.

The nature of this research is descriptive-exploratory, since it intends to constitute a preliminary approximation to this unknown field of study. In addition, this study does not pursue statistical extrapolation, since its objective, instead of being statistical inference, is the non-parametric exploration of the phenomenon. Therefore, the methodology of the study was the expert consultation and the technique selected to obtain the primary information was the structured interview through a questionnaire (survey).

4 RESULTS AND DISCUSSION

Results obtained in the present study are involved in the three chief sections of the designed survey:

4.1 General aspects

One of the aims of this sub-section was to analyse the risk perception that experts have about the future of the reuse of treated wastewaters due to the presence of EPs.

Regarding the risk perception about EPs issues (Fig. 3(a)), most responses of the surveyed experts were classified in the category of "High" with a 40.7% following by the "Medium" (26.5%) and the "Very High" (24.8%) categories. This inform that a 92% of the 114 experts consider as relevant these problems. In particular, conversion of the ordinal levels, achieved in this question, into numerical scale positions denotes that the average score of this variable (risk perception) is 3.87/5, obtaining therefore the highest value of the variables subset which integrates this section called General aspects (Fig. 3(a)). Respect the adaptation of the treatment systems to meet the requirement of new regulations regarding EPs (Fig. 3(b)), most experts assigned values of "Low" (38.9%), "Very low" (23.0%), etc. Likewise, systems preparation achieved 2.26/5.

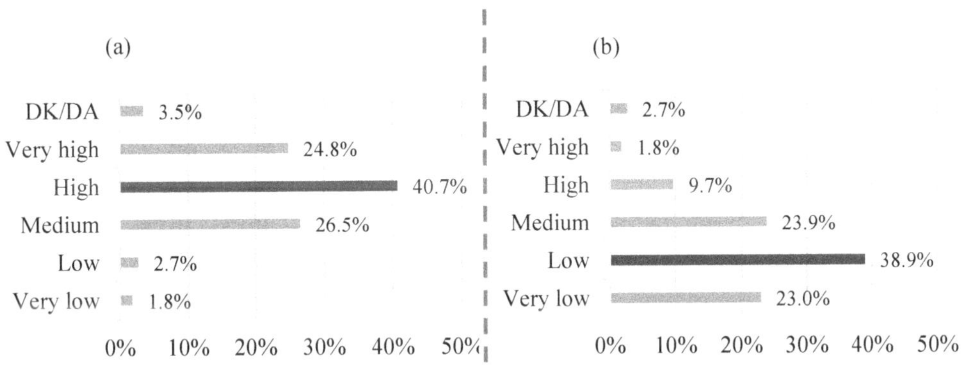

Figure 3: (a) Risk perceived about EPs; (b) Adaptation of the treatment systems.

At the same time, application of new regulations related with EPs issues is considered as "Positive" (31.0%) and "Very Positive" (23.9%). Therefore, this variable (impact of new regulations) achieved a great average score (3.55/5, Fig. 4(a)). According with results obtained in the application of new normative, most experts indicated the relevance and needs for new techniques which can determine new, and probably lower, concentrations of EPs fixed by the planned regulations (Fig. 4(b)). In particular, the "Development of analysis techniques" variable acquired an average score of 3.73/5.

The assess related with the cost of new equipment, or certain improvements in the existing equipment, generated rejection between the surveyed experts being the "Regular" and "Negative" categories the most selected (Fig. 5(a)). This variable obtained an average score of 2.89/5. On the other hand, experts estimated improvements of the EPs scientific knowledge as essential being the "Very high" category the most selected with a 30.1% (Fig. 5(b)). In particular, this variable gained an average score of 3.78/5.

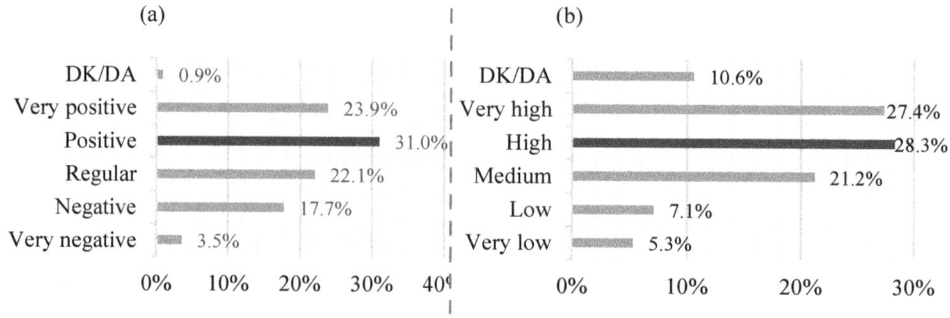

Figure 4: (a) Impact of new regulations; (b) Development of analysis techniques.

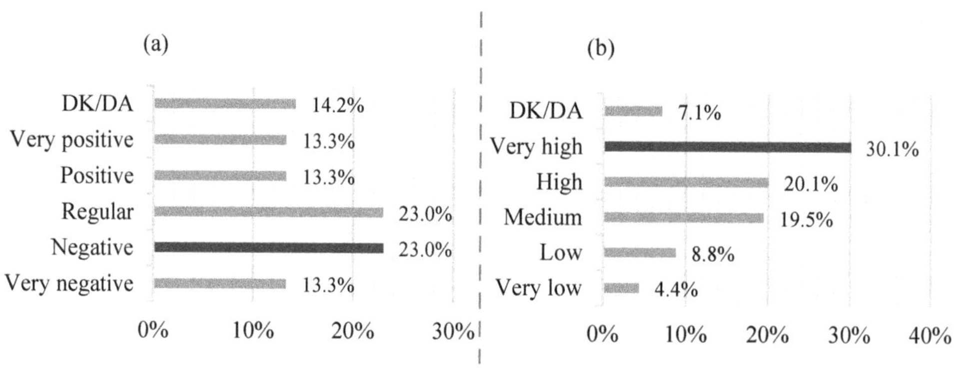

Figure 5: (a) Cost of analysis equipment; (b) Training and scientific knowledge.

4.2 Technologies

One of the aims of this sub-section was to assess the perception that the experts have about the effectiveness of the different processes to remove EP: membrane processes (Ultrafiltration and Reverse Osmosis), Advanced Oxidation Processes, Electrochemical Processes and Adsorption. The responses of the surveyed experts are showed in Fig. 6.

Regarding the membrane process Ultrafiltration (Fig. 6(a)), most experts classified the effectiveness of this process to remove EPs in the category "High" (27.4%) and others in "Very high" (11.5%). The 24.8% of the participants classified the effectiveness as "Medium", and only a 18.5% of them choose the category "Low" (15%) and "Very low" (3.5%). This inform that a 63.7% of the 114 experts consider this technology can efficiently remove EPs from wastewaters. The average score of this variable (effectiveness of Ultrafiltration to remove EPs) into numerical scale is 3.34/5.

With respect to Reverse Osmosis (Fig. 6(b)), most experts classified the effectiveness of this process to remove EPs in the category "High" (31.9%) and others in "Very high" (23.0%). The 17.7% of the participants classified the effectiveness as "Medium", and only a 9.7% of them choose the category "Low" (8.8%) and "Very low" (0.9%). This inform that a 72.6% of the 114 experts consider this technology can efficiently remove EPs from wastewaters. The average score of this variable (effectiveness of Reverse Osmosis to remove

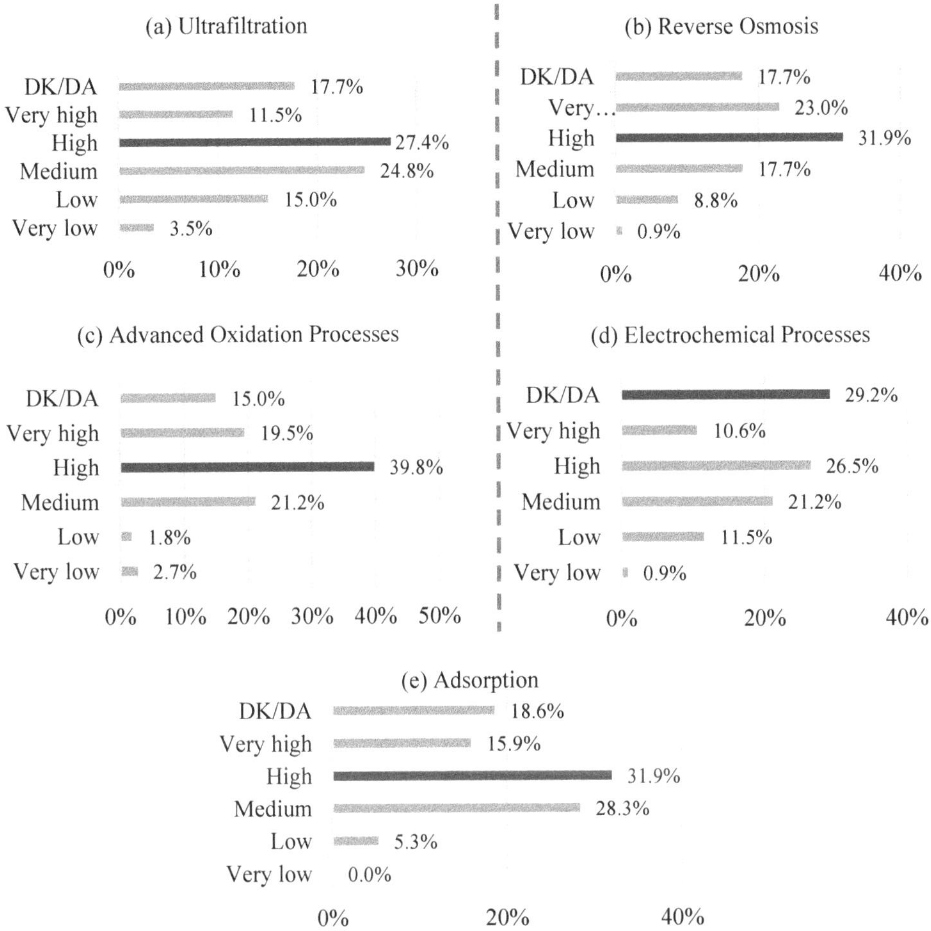

Figure 6: Technologies to remove EPs from wastewaters.

EPs) into numerical scale is 3.82/5. Comparing the average scores for membrane processes, experts considered more efficient the Reverse Osmosis process.

In relation to Advanced Oxidation Processes (Fig. 6(c)), most experts classified the effectiveness of this process to remove EPs in the category "High" (39.8%) and others in "Very high" (19.5%). The 22.2% of the participants classified the effectiveness as "Medium", and only a 4.5% of them choose the category "Low" (1.8%) and "Very low" (2.7%). This inform that the 80.5% of the 114 experts consider this technology can efficiently remove EPs from wastewaters. The average score of this variable (effectiveness of Advanced Oxidation Processes to remove EPs) into numerical scale is 3.84/5.

Regarding Electrochemical Processes (Fig. 6(d)), most experts classified the effectiveness of this process to remove EPs in the category "High" (26.5%) and others in "Very high" (10.6%). The 21.2% of the participants classified the effectiveness as "Medium", and only a 12.4% of them choose the category "Low" (11.5%) and "Very low" (0.9%). This inform that a 58.3% of the 114 experts consider this technology can efficiently remove EPs from wastewaters. The average score of this variable (effectiveness of Reverse Osmosis to remove EPs) into numerical scale is 3.49/5. It is worth noting that this technology was the less known by the experts (29.2% of the responses in category DK/DA).

Concerning Adsorption (Fig. 6(e)), most experts classified the effectiveness of this process to remove EPs in the category "High" (31.9%) and others in "Very high" (15.9%). The 28.3% of the participants classified the effectiveness as "Medium", and only a 5.3% of them choose the category "Low" (5.3%) and "Very low" (0.0%). This inform that a 76.1% of the 114 experts consider this technology can efficiently remove EPs from wastewaters. The average score of this variable (effectiveness of Adsorption to remove EPs) into numerical scale is 3.72/5.

Considering the opinion of experts (average scores), the processes can be ordered from more to less effectiveness to remove EPs as: Advanced Oxidation Processes, Reverse Osmosis, Adsorption, Electrochemical Processes and Ultrafiltration.

Other aim of this sub-section was to assess the perception that the experts have about the need for the combination of technologies to obtain effluents without EPs. The responses of the surveyed experts are showed in Fig. 7.

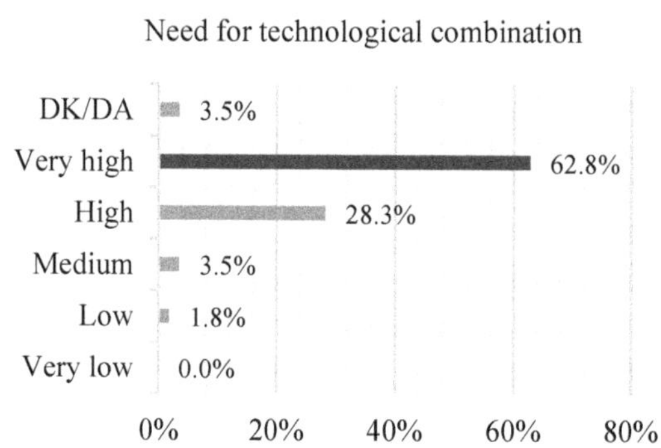

Figure 7: Combination of technologies to obtain effluents without EPs.

As it was previously commented, there were different opinions about the effectiveness of each individual process (Ultrafiltration, Reverse Osmosis, Advanced Oxidation Processes, Electrochemical Processes and Adsorption) to remove EPs from wastewaters. However, it dares highlighting that most experts (91.1% of the 114 experts) agreed on the need to combine those technologies to efficiently remove EPs, as their responses confirmed: "High" (28.3%) and "Very high" (62.8%). Only the 5.3% of the participants classified this need as "Medium" (3.5%), "Low" (1.8%) and "Very low" (0.0%). The average score of this variable (need for technological combination to remove EPs) into numerical scale is 4.58/5.

4.3 Social, economic and environmental impact of EPs management

The aim of this sub-section was to analyse the impact that the combination of technologies for the removal of EPs would have on the following social, economic and environmental spheres.

In relation to the impact on public health, the most repeated category was "very positive", with 55.8% of the answers. This category and the option "positive" combine more than 85%

of the total responses. Thus, the average score of this variable is 4.44/5, the highest value of the subset of variables on the impact generated by the elimination of EPs. This data illustrates the strong relationship between EPs and health. The effect of EPs on human health is the main cause of concern for expert users (Fig. 8(a)).

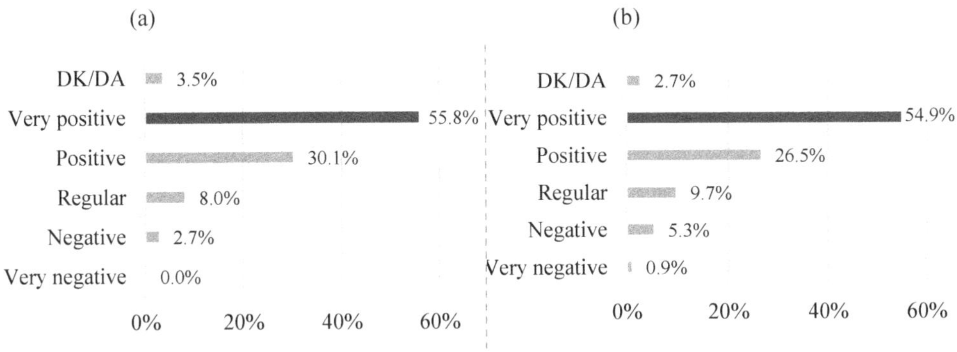

Figure 8: (a) Impact of EPs on public health; (b) Impact of EPs on environmental sustainability.

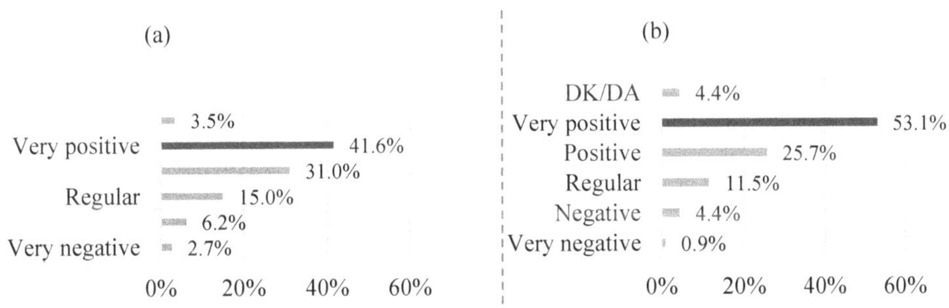

Figure 9: (a) Impact on water resources management; (b) Impact on social trust.

Regarding the impact on environmental sustainability (Fig. 8(b)), the responses are also homogeneous. More than half of the people who responded to the survey considered that the impact would be "very positive." The categories "regular" and "negative" represent residual percentages. The average score of this variable is 4.33/5. A greater control of the discharges, materialized through an adequate elimination of EPs, could generate positive impacts at the environmental level, such as the increase in the quality of the ecological flow of the rivers and their corresponding ecosystem benefits.

On the other hand, in relation to the impact that the elimination of EPs (Fig. 9(a)) would have on the management of water resources, the data are also quite positive. The average score of this variable, however, is the lowest of this sub-set of variables, with a value of 4.06/5. Despite being the lowest average, it exceeds 4, so it is a positive assessment. The EPs, as has been verified with the data of other variables, are a very important factor to understand the assessment and the future of water reuse. Therefore, its elimination could mean an increase in the consumption of reclaimed water (Fig. 9(a)).

In terms of social trust, the data is very similar, although slightly better than the previous variable. The average score is 4.31/5. Therefore, the categories "very positive" and "positive" concentrate almost 80% of the total responses. The concerns of expert users about the effects of reclaimed water on human health are equivalent to those of citizens in general (March et al. [17]). A EPs-free water could generate greater social trust over non-conventional water resources (water reuse and desalination, Fig. 9(b)).

The second lowest average score of this sub-set of variables was the impact on the laboral market (Fig. 10(a)), with a value of 4.11/5. In this case, the most repeated category is "positive", with 40.7% of the answers. This category and the option "very positive" concentrate more than 75% of the answers. The overall assessment of this variable is significantly positive. Technological advances and investment in projects related to the elimination of EPs could stimulate some economic sectors and facilitate the financing of the scientific staff of some research centers.

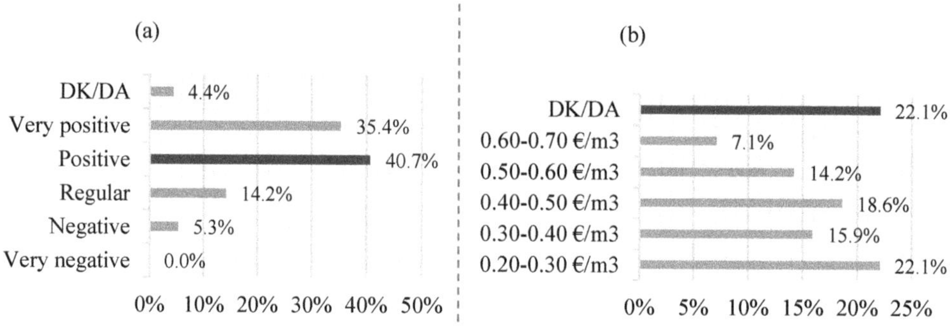

Figure 10: (a) Impact on laboral market; (b) Acceptable price.

An economic aspect that is interesting to analyse is the acceptable price for EPs-free water (Fig. 10(b)). It is worth noting the high level of ignorance about the price of water. 22.1% of people did not know the price. This option is the most repeated category. Secondly, the acceptable price of EPs-free water that users indicated is "very low". 22.1% of people considered that the most acceptable price range is "0.20-0.30 €/m^3", the lowest level proposed. This information is consistent with the data of the variable on the willingness of users to pay more for EPs-free water (Fig. 11). 42.5% of users indicated they were willing to assume an increase in the price as long as it was moderate. About 18% of the experts indicated that they are willing to assume a higher price if the increase is minimal.

5 CONCLUSIONS

Main conclusions of the present study are listed below:

- To sum up, the section called General Aspects presented main results of the study related with the risk perception, adaptation systems, impact of new regulations, development of analysis techniques, cost of analysis equipment, training and scientific knowledge. It should be noted that the risk perception and the improvement of the scientific knowledge were the variables considered the most relevant of this section. Obviously, these are variables closely related since the high-risk perception can be considered the reason by experts estimated improvement of the scientific knowledge as essential.

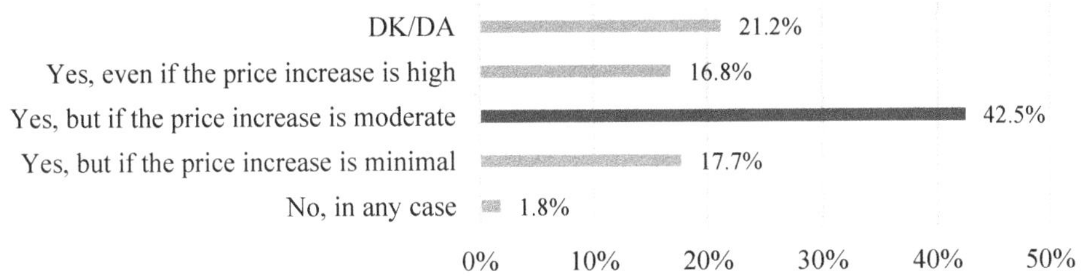

Figure 11: Propension to pay more.

- Considering the opinion of experts (average scores), the processes can be ordered from more to less effectiveness to remove EPs as: Advanced Oxidation Processes, Reverse Osmosis, Adsorption, Electrochemical Processes and Ultrafiltration. It dares saying that Electrochemical Processes was the less known technology by the experts.
- It is worth noting that most experts agreed on the need to combine these technologies to efficiently remove EPs from wastewaters.
- In general, users consider that the elimination of EPs can stimulate positive impacts at a social, economic and environmental level, highlighting the perception of impact in the field of public health. The effect of EPs on human health could be considered the main concern of users.
- Although experts recognize many benefits from the elimination of EPs, they are not willing to assume a significant increase in costs. There is a disproportionate relationship between concerns, expectations and willingness to adapt to new economic realities. This constitutes a factor of uncertainty and vulnerability that affects the current wastewater management and treatment systems.

ACKNOWLEDGEMENTS

This projcct was produced under the co-finance of the European financial instrument for the Environment (LIFE) programme during the implementation of the project "LIFE EMPORE" (LIFE15 ENV/ES/000598). Pablo Aznar-Crespo and Antonio Jodar-Abellan acknowledge financial support received from the Spanish FPU scholarship for the training of university teachers. In the same way, this work has been conducted within the *Cátedra del Agua* of the University of Alicante (https://catedradelaguaua.org/). Finally, the authors acknowledge the reviewers of the manuscript whose comments contributed greatly to improve this paper.

REFERENCES

[1] Directive 2000/60/EC of the European Parliament and of the Council of 23 October 2000, Establishing a framework for Community action in the field of water policy.

[2] Directive 2013/39/EU of the European Parliament and of the Council, of 12 August 2013, Amending Directives 2000/60/EC and 2008/105/EC as regards priority substances in the field of water policy.

[3] Directive 91/271/EEC. Council Directive of 21 May 1991, Concerning urban waste water treatment.

[4] Directive 98/15/EC. Commission Directive 98/15/EC of 27 February 1998, Amending Council Directive 91/271/EEC with respect to certain requirements established in Annex I thereof (Text with EEA relevance).
[5] BIO Deloitte, Optimising water reuse in the EU. Final report for the European Commission (DG ENV), Part I. In collaboration with ICF and Cranfield University, 2015.
[6] Confederación Hidrográfica del Segura (CHS), Plan Hidrológico de la Demarcación del Segura, Memoria, Ministry of Agriculture and Environment, p. 1–762, 2015.
[7] Gomariz-Castillo, F., Alonso-Sarria, F. & Cabezas-Calvo-Rubio, F., Calibration and spatial modelling of daily ET0 in semiarid areas using Hargreaves equation. *Earth Science Informatics*, **11**(3), pp. 325–340, 2017.
[8] Ruíz-Álvarez, M., Gomariz-Castillo, F. & Alonso, F., Análisis espacio-temporal de los cambios previstos en la evapotranspiración de referencia durante el siglo XXI en la Demarcación Hidrográfica del Segura (España) a partir de los modelos MPEH5 y MPEH5C. *Revista de Geografía Norte Grande*, **71**, pp. 35–58, 2018.
[9] Albaladejo-García, J.A., Martínez-Paz, J.M. & Colino, J., Evaluación financiera de la viabilidad del uso de agua desalada en la agricultura de invernadero del Campo de Níjar (Almería, España). *Información técnica económica agraria*, 114(**4**), pp. 398–414. 2018.
[10] Grindlay-Moreno, A.L. & Lizárraga-Mollinedo, C., Regadío y territorio en la Región de Murcia: evolución y perspectivas de futuro. *Ciudad y Territorio*, **172**, pp. 281–298, 2012.
[11] Custodio, E. et al., Groundwater intensive use and mining in south-eastern peninsular Spain: Hydrogeological, economic and social aspects. *Science of the Total Environment*, **559**(1), pp. 302–316, 2016.
[12] Melgarejo-Moreno, J., López-Ortiz, M.I. & Fernández-Aracil, P., Water distribution management in South-East Spain: A guaranteed system in a context of scarce resources. *Science of the Total Environment*, **648**, pp. 1384–1393, 2019.
[13] Molina-Giménez, A. & Melgarejo-Moreno, J., Water policy in Spain: seeking a balance between transfers, desalination and wastewater reuse. *International Journal of Water Resources Development*, **32**(5), pp. 781–798, 2015.
[14] Zarzo, D. & Prats, D., Desalination and energy consumption. What can we expect in the near future? *Desalination*. **427**, pp. 1–9, 2018.
[15] Prats-Rico, D., Reuse of Purified Regenerated Water Worldwide: Analyzes and Projections. *Water and landscape*. **8**, pp. 10–21, 2016.
[16] SPSS code: Statistical Product and Service Solutions, www.spss.com.hk/. Accessed on: 22 Nov. 2018.
[17] March, H., Hernández, M. & Saurí, D., Percepción de recursos convencionales y no convencionales en áreas sujetas a estrés hídrico: el caso de Alicante. *Revista de Geografía Norte Grande*, **60**, pp. 153–172, 2015.

ASSESSMENT OF DAILY INTAKE OF ARSENIC AND ASSOCIATED HEALTH RISKS FOR CHILDREN

IMOKHAI T. TENEBE[1,2], PRAISEGOD C. EMENIKE[1], DIWA I. DANIEL[1], AIKUOLA OLUMUYIWA[3], CHUKWUKA CHIBUIKE DANIEL[1], OLUNUGA OLUWATOSIN GIDEON[1], ABRAHAM CHUKWUEBUKA[1], UMEH CHUBUIKE[1] & EGBU-ETU EMMANUEL ETU[4]
[1]Department of Civil Engineering, Covenant University, Nigeria
[2]Ingram School of Engineering, Texas State University, USA
[3]Federal Polytechnic Ilaro, Nigeria
[4]Department of Industrial and Systems Engineering, Wayne State University, USA

ABSTRACT
Surface water is a major source of heavy metal contamination due to land use variabilities and can impair the health status of humans especially children when consumed. This study investigated the arsenic concentration in surface water of River Balogun in south-west Nigeria. Fifty-one samples were collected from 17 points along the river and were examined using inductively coupled plasma optical emission spectrophotometer (ICP–OES) for arsenic concentration levels and the potential of a risk due to prolonged exposures in children using average daily dose, and hazard quotient (HQ) indices were estimated. From this study, it revealed that the mean daily intake concentrations of arsenic for children were within the range of 0.0034–0.0068 mg/l and these values were below the standards for the World Health Organization. However, long term exposures could be injurious to the health of children as the HQ values were greater than one for all 51 samples. Therefore, continuous human health risk assessment, sensitization on the adverse effects on the overuse of agrochemicals in farms surrounding the rivers, intensified ecosystem services as well as best management practices for improvement of surface water in the region need to be adopted. Furthermore, elaborating on the dangers of high arsenic in water can also be a strategy in preventing children and other population groups from consuming contaminated surface waters in that vicinity.
Keywords: risk assessment, arsenic, pollution, heavy metals, dispersion, hazard quotient.

1 INTRODUCTION

Access to safe water is germane and this has led to the Sustainable Development Goals (SDGs) Target 6, Goal 6 which aimed at providing cheap, safe and less polluted water for all [1], [2]. Rain-harvested, surface, sachet waters are some of the sources where water can be assessed. However, getting this water free from chemical or bacteriological pollution is still very difficult especially in developing nations. Several reports have revealed the chemical toxicity levels of heavy metals at significant levels in some of these water sources in an alarming rate [3]–[5]. The reasons for this pollution levels may be due to land use and poor hygiene as there is a strong significant relationship between the environment and the pollution status of water bodies or sources [6]–[8] as well as lack of consciousness and convincing awareness from the inhabitants and relevant agencies. These metals can cause several health effects that can reduce human efficiency at the least and may result in death at worst. Some of these adverse effects have been well reported in the literature [9]–[11]. Arsenic (As), which is one of these heavy metals which are found naturally is of great concern for public health because of its effects on the human health such as dermal and cardiovascular diseases, skin and bladder cancer, diabetes mellitus among others [12], [13]. Its presence in food proves that it is usually accumulated from the environment [14]. In groundwater, As is found mainly in inorganic forms [15] which are considered much more toxic than the organic one [16]. It is estimated that close to 140 million people globally are exposed to inorganic As which are primarily ingested by consuming unregulated contaminated water [17]. Thus, As

WIT Transactions on Ecology and the Environment, Vol 234, © 2019 WIT Press
www.witpress.com, ISSN 1743-3541 (on-line)
doi:10.2495/RBM190191

in the drinking water sources, be it ground or surface waters, is of important concern due to its associated health risk [18]. In the light of this, this study investigated the pollution level of As in surface water of River Balogun, located in Ota. This river is surrounded by industries and it is well utilized by the inhabitants for domestic and recreational purposes. Due to the open nature of the river, it is mostly used as a carrier of unwanted substances with the believe that with dilution and dispersion, these inherent pollutants concentration may be reduced [19]. Therefore, in this paper, we examine the health risk linked with surface water daily consumption of As to children.

2 METHOD AND MATERIAL

2.1 Study area

River Balogun situated around Ado-Odo/Ota local government area in south-west Nigeria, with an area of 1,460 km^2. It lies between latitude 6°40'58.52"N–6°41'23.92"N and longitude 3°8'53.87"E– 3°8'57.86"E and is a tributary of the River Atuwara. The study area is densely populated with informal settlements situated close to the study area. River Balogun traverses across villages, providing the needed water and economic activities to the surrounding villages such as dredging, irrigation farming and fishing. River Balogun's water is essentially important for the ecosystem, health and sustained economic activities such as agriculture and supply of drinking water to people, which requires continuous environmental surveillance and proper industrial waste management.

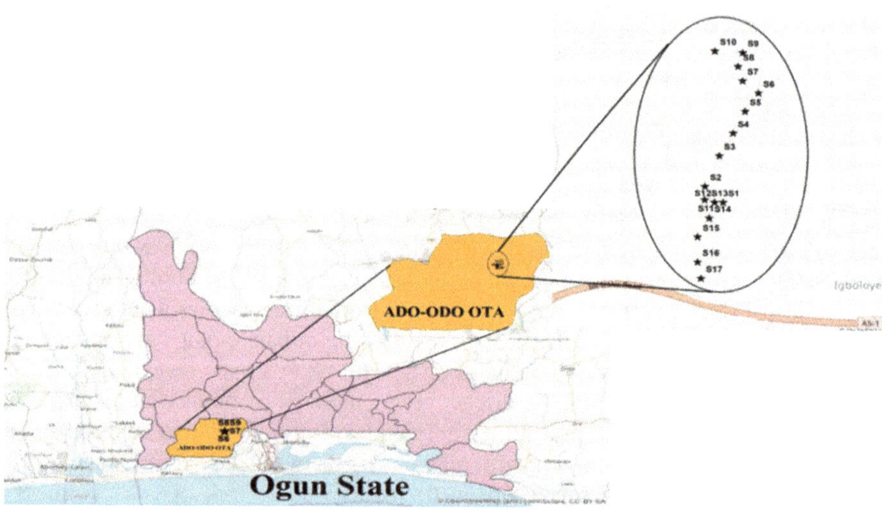

Figure 1: A map showing River Balogun at Ota, Ogun State [10].

2.2 Sampling collection and procedures

Fifty-one (51) water samples from seventeen (17) stations were collected from the river using a well labelled High density polyethene bottle. The samples collected were iced in a cooler at 4°C. Thereafter, these samples were sent to the International Institute for Tropical Agriculture, Ibadan, Nigeria, for chemical analysis. The samples were analysed for As using Inductively Coupled Plasma–Optical Emission Spectrometry (ICP–OES).

2.3 Health risk assessment

Generally, risk assessment is a function of hazard while exposure is the process of estimating the probability of occurrence of an event and the probable magnitude of adverse health effects produced over a specified time period [20]. Human health risk assessment is considered as the characterization of the potential adverse health effects of humans due to exposure to environmental hazards [21]. This paper estimated the average daily dose (ADD) expected for the intake of As using the following eqn (1), while cancer risk and hazard quotient were estimated using (2) and (3) respectively

$$ADD = C \times IR \times ED \times EF/BW \times AT, \tag{1}$$

where ADD is the average daily dose (mg/kg-day), C is the concentration of the heavy metal in stream water (mg/l), IR is the ingestion rate per unit time (L/day), ED is the exposure duration (years), EF is the exposure frequency (days/year), BW is body weight (kg), AT is the average time (years), with their values obtained from literatures [22]–[24].

The cancer risk was calculated as outlined by [21]:

$$\text{Cancer risk} = ADD \times CSF, \tag{2}$$

where ADD = average daily doses (mg/kg-day); CSF = cancer slope factor (mg/kg-day).

The calculated ADD and cancer risk are as presented in Fig. 2.

$$HQ_{ingestion} = \frac{ADD_{ingestion}}{RfD_{ingestion}}. \tag{3}$$

3 RESULTS AND DISCUSSION

The ADD values for As are presented in Fig. 2. For children, the daily intake for As was between range of 0.0034–0.0068 mg/kg-day. According to the results, the level of As, 6.799 E-03 mg/kg-day for children in Station 1 and Station 2 respectively appeared high, but the highest values of 8.500 E-03 were recorded at stations 11, 12, 13,15, 16 and 17 for

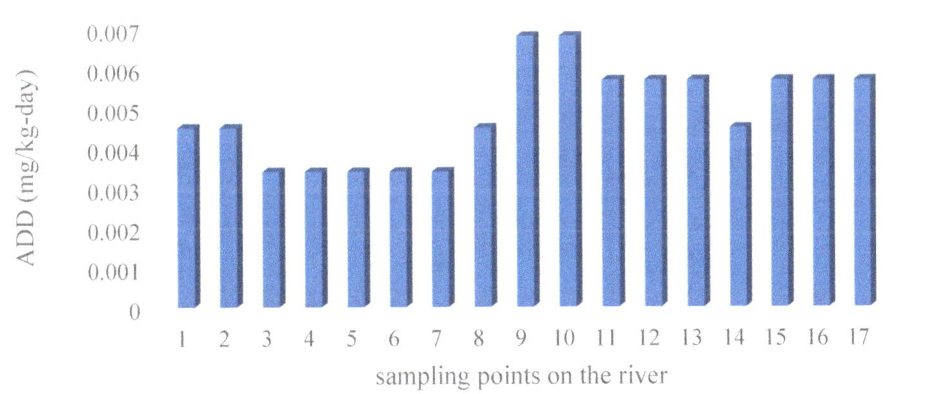

Figure 2: Graphical representation of average daily intake of As in children.

children. Furthermore, due to the measured results and importance of the ELCR (excess lifetime cancer risk) level – which is the threat to human health, it becomes highly important that drinking water control and quality in that region is a necessity specially to protect the children. In addition, results revealed that River Balogun water may pose a high risk, when compared with the water quality guidelines of USEPA [25], which is fixed at 10 μg/l. Consequently, cancer risk assessment of the study area suggested that there is high health risk due to the As contamination. As was reported to be detrimental to the central nervous system and cognitive development in children [26]. It was also found to accumulate in fingernails and hair [27] which can later find its way into our body. Therefore, it implied that using this water either for ingestion or recreation is dangerous.

Fig. 3 shows the HQ of As consumption for children in River Balogun at different sampling stations. The HQ values recorded ranged from 11.33–22.67 mg/l. The highest level was at Stations 9 and 10 which recorded 22.67 mg/l as against World Health Organization [28] standard of 1.00 mg/l. This result indicates that children ingest water that is significantly polluted with As (about 22 times more than the threshold value) and appeared to be at high risk for carcinogenic effects in the long run. In other stations, high As levels were equally recorded. Specifically, stations 11–16, with values as 19 mg/l while 15 mg/l was recorded in stations 1 and 2. Furthermore, 11 mg/l were recorded at stations 3, 4, 5, 6 and 7 as well. This may be due to over application of fertilizers and poor agricultural practices within the study area as reported in similar studies elsewhere [29], [30]. These results agreed with the findings of [31] in Taiwan where the HQ values were 19.3 mg/g revealing threat to life of the population. Subsequently, when children ingest water polluted by As, they become more exposed to all kinds of cancers especially cancer related to the skin, kidney and bladder which are classified as carcinomas [32]. As also known to speed up cell division which leads to all kinds of health complications in human. In addition, retardation of time required for enzymes to repair DNA damage during DNA replication will occur and the likelihood of genetic misleading may be the consequence.

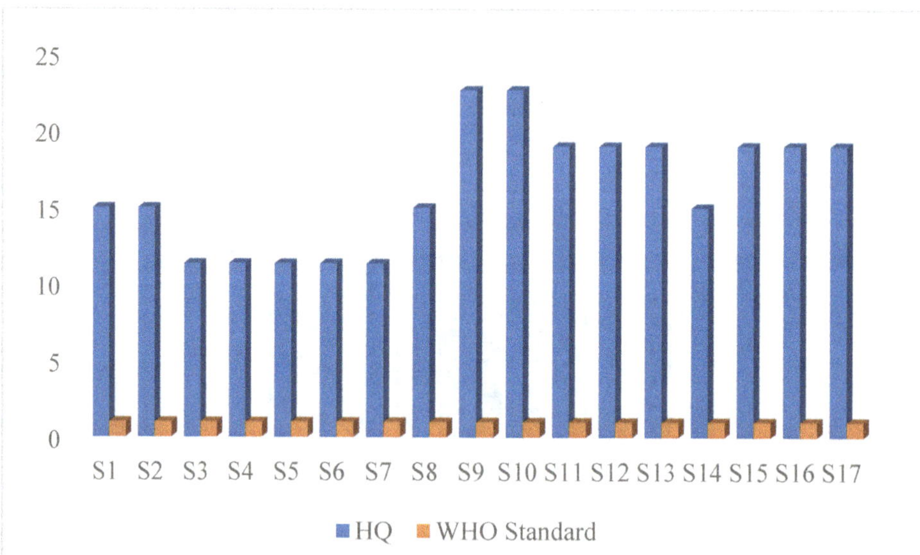

Figure 3: Graphical representation of the hazard quotient (As for children) against the World Health Organization standard.

4 CONCLUSION

The ADD values and cancer risks for As were highest in Stations 11, 12, 13, 15, 16, and 17 for children, which can pose a threat to human life. Due to the carcinogenic tendency of As over long periods of exposure and the risk to life and health, it is important to control the presence of As in the river Balogun as the present status showed a significant health risk to children. Prevention of exposure to As in drinking water is a major strategy for reducing the incidence of cancer due to As ingestion. In addition, regular monitoring of contamination levels of heavy metals and imbibing good sanitation practises and convincing awareness to inhabitants on the dangers of As pollution to children would boost general water quality of the river as well as other water sources. In addition, improving ecosystem services around rivers in general by creating wetlands or riparian buffers which should prevent agricultural runoff from gaining easy entrance into the river would serve as some kind of best management practices to be swiftly adopted.

ACKNOWLEDGEMENTS

We are very grateful to Covenant University and Texas State University for their sponsorship.

REFERENCES

[1] UNDESA, Sustainable development goals. https://sustainabledevelopment.un.org/sdgsproposal. Accessed on: 7 Nov. 2015.

[2] Emenike, C.P., Tenebe, I.T., Omole, D.O., Ndambuki, J.M., Ogbiye, A.S. & Sojobi, A., Application of water recovery option for agricultural use in developing countries: Case study of a Nigerian community. *Conference on International Research on Food Security, Natural Resource Management and Rural Development*, 2015.

[3] Chinedu, S., Nwinyi, O., Oluwadamisi, A. & Eze, V., Assessment of water quality in Canaanland, Ota, Southwest Nigeria. *Agriculture and Biology Journal of North America*, **2**(4), pp. 577–583, 2011.

[4] Anake, W., Benson, N., Akinsiku, A., Ehi-Eromosele, C. & Adeniyi, I., Assessment of trace metals in drinking water and groundwater sources in Ota, Nigeria. *International Journal of Scientific and Research Publications*, **4**, pp. 1–4, 2014.

[5] Emenike, P.C., Tenebe, T.I., Omeje, M. & Osinubi, D.S., Health risk assessment of heavy metal variability in sachet water sold in Ado-Odo Ota, South-Western Nigeria. *Environ. Monit. Assess.*, **189**, p. 480, 2017. DOI: 10.1007/s10661-017-6180-3.

[6] Tenebe, I.T., Emenike, C.P., Ogbiye, A.S., Ngene, B.U., Omeje, M. & Olatunji, O.O., A laboratory assessment of the effect of varying roughness on dissolved oxygen using error correction method. *Cogent Engineering*, pp. 1–11, 2018.

[7] Tenebe, I.T., Ogbiye, A.S., Omole, D.O. & Emenike, P.C., Estimation of longitudinal dispersion co-efficient: A review. *Cogent Engineering*, **3**, 2016. DOI: 10.1080/23311916.

[8] Emenike, P., Omole, D., Ngene, B. & Tenebe, I., Potentiality of agricultural adsorbent for the sequestering of metal ions from wastewater. *Global Journal Environmental Science and Management*, **2**, pp. 411–442, 2016.

[9] Muhammad, N., Bangush, M. & Khan, T., Microbial contamination in well water of temporary arranged camps: A health risk in northern Pakistan. *Water Quality, Exposure and Health,* **4**(4), pp. 209–215, 2012. DOI: 10.1007/s12403-012-0080-0.

[10] Tenebe, I.T., Emenike, C.P. & Chukwuka, D., Prevalence of heavy metals and computation of its associated risk in surface water consumed in Ado-Odo Ota, South-West Nigeria. *Human and Ecological Risk Assessment: An International Journal*, pp. 1–23, 2018. DOI: 10.1080/10807039.2018.1454824.

[11] Emenike, P.C., Tenebe, I.T. & Jarvis, P., Fluoride contamination in groundwater sources in Southwestern Nigeria: Assessment using multivariate statistical approach and human health risk. *Ecotoxicol. Environ. Saf.*, 2018. DOI: 10.1016/j.envsoft.2004.03.001.
[12] Eblin, K.E. et al., Arsenite and monomethylarsonous acid generate oxidative stress response in human bladder cell culture. *Toxicology and Applied Pharmacology*, **217**(1), pp. 7–14, 2006.
[13] Sun, Q. et al., Improved diffusive gradients in thin films (DGT) measurement of total dissolved inorganic arsenic in waters and soils using a hydrous zirconium oxide binding layer. *Analytical Chemistry*, **86**(6), pp. 3060–3067, 2014.
[14] Roychowdhury, T., Uchino, T., Tokunaga, H. & Ando, M., Survey of arsenic in food composites from an arsenic-affected area of West Bengal, India. *Food and Chemical Toxicology*, **40**, pp. 1611–1621, 2002.
[15] Tuzen, M., Citak, D., Mendil, D. & Soylak, M., Arsenic speciation in natural water samples by coprecipitation-hydride generation atomic absorption spectrometry combination. *Talanta*, **78**(1), pp. 52–56, 2009.
[16] Brahman, K.D., Kazi, T.G., Afridi, H.I., Naseem, S., Arain, S.S. & Ullah, N., Evaluation of high levels of fluoride, arsenic species and other physicochemical parameters in underground water of two sub districts of Tharparkar, Pakistan: A multivariate study. *Water Research*, **47**(3), pp. 1005–1020, 2013.
[17] States, J.C., Barchowsky, A., Cartwright, I.L., Reichard, J.F., Futscher, B.W. & Lantz, R.C., Arsenic toxicology: translating between experimental models and human pathology. *Environ. Health Perspect.*, 2011. DOI: 10.1289/ehp.1103441.
[18] Sorg, T.J., Chen, A.S.C. & Wang, L., Arsenic species in drinking water wells in the USA with high arsenic concentrations. *Water Research*, **48**(1), pp. 156–169, 2014.
[19] Tenebe, I.T., Ogbiye, A.S., Omole, D.O. & Emenike, P.C., Modelling and sensitivity analysis of varying roughness effect on dispersion coefficient: A laboratory study. *Desalination and Water Treatment*, **87**, pp. 1–7, 2017. DOI: 10.5004/dwt.2017.21298.
[20] Wongsasuluk, P., Chotpantarat, S., Siriwong, W. & Robson, M., Heavy metal contamination and human health risk assessment in drinking water from shallow groundwater wells in an agricultural area in Ubon Ratchathani province, Thailand. *Environ. Geochem. Health*, **36**, pp. 169–182, 2014.
[21] USEPA, Waste and cleanup risk assessment. www2.epa.gov/risk/waste-and-cleanup-risk-assessment. Accessed on: Dec. 2017.
[22] USEPA, *United States Environmental Agency Risk Assessment Guidance for Superfund, Volume 1: Human Health Evaluation Manual (Part A)*, Interim Final Office of Emergency and Remedial Response, EPA/540/7-89/002, 1989.
[23] USEPA, *United States Environmental Protection Agency, Human Health Evaluation Manual, Supplemental Guidance, Standard Default Exposure Factors,* OSWER Dir. 9285, 6–03, 1991.
[24] Van Graan, A.E., Bopapa, M., Phooko, D., Bourne, L. & Wright, H.H., Drink lots of clean safe water: A food-based dietary guideline for South Africa. *S. Afr. J. Clin. Nutr.*, **26**(3), pp. S77–S86, 2013.
[25] USEPA, Integrated risk information system (IRIS), United States Environmental Protection Agency. www.Epa.Gov/irrris/index.infml.
[26] Rosado, J.L. et al., Arsenic exposure and cognitive performance in Mexican schoolchildren. *Environ. Health Perspect.*, **115**(9), pp. 1371–1375, 2007.

[27] Choong, T.S.Y., Chuah, T.G., Robiah, Y., Koay, F.L.G., Azni, I., Arsenic toxicity, health hazards and removal techniques from water: An overview. *Desalination*, **217**, pp. 139–166, 2007.

[28] World Health Organization, *Guideline for Drinking-water Quality, 3rd ed. incorporating 1st and 2nd Agenda, Vol. 1: Recommendations*, Geneva, 668 pp., 2008.

[29] Tsuda, T. et al., Historical cohort studies in 3 arsenic poisoning areas in Japan. *Applied Organometallic Chemistry,* **6**, pp. 333–341, 1992.

[30] Christ, O., Charalambous, M., Aletrari, C., Nicolaidou Kanari, M., Petronda, P., Ward, N.I., Arsenic concentrations in ground waters of Cyprus. *J. Hydrol.*, **468**(469), pp. 94–100, 2012.

[31] Han, B.-C., Jeng, W.-L., Hung, T.-C., Ling, Y.-C., Shieh, M.-J. & Chien, L.-C., Estimation of metal and organochlorine pesticide exposures and potential health threat by consumption of oysters in Taiwan. *Environ. Pollut.*, **109**, pp. 147–156, 2000.

[32] Lamm, S.H., Engel, A., Penn, C.A., Chen, R. & Feinleib, M., Arsenic cancer risk confounder in Southwest Taiwan data set. *Environ. Health Perspect.*, **114**, pp. 1077–1082, 2006.

www.witpress.com, ISSN 1743-3541 (on-line)

LONGITUDINAL DISPERSION PREDICTION OF RIVER BALOGUN USING CONSTANT DISTANCE AND TIME APPROACH AT DIFFERENT SAMPLING TIME INTERVALS

IMOKHAI T. TENEBE[1], OGBIYE A. SAMUEL[1], DAVID O. OMOLE[1], PRAISEGOD C. EMENIKE[1], IREWOLE B. ONIEMAYIN[1], DIWA I. DANIEL[1] & OMEJE MAXWELL[2]
[1]Department of Civil Engineering, Covenant University, Nigeria
[2]Department of Physics, Covenant University, Nigeria

ABSTRACT

The estimation of dispersion number or its coefficients can be used to determine the degradation capacity of rivers. This can be determined using tracer studies with the different methods found in the literature. However, for this study, the constant distance and time method was used at two different sampling time intervals to determine its value as well as the implication of being subjective in sampling time interval selection. This study showed that the dispersion number obtained from River Balogun using the Levenspiel and Smith approach at the 3 mins time interval (d=0.037152) and 5 mins interval (d=0.0214) varied by a factor of 1.76. This implies that the dispersion number could be underestimated or overestimated. Furthermore, the first sampling interval required 1.54 times more tracer concentration data than the second sampling time interval thereby incurring more cost, time and labour. Therefore, it is suggested that more sustainable approaches are developed and practised that would use limited tracer data collection for calculation as well as limit subjectivity in sampling time selection, yet give closely precise values of dispersion number.

Keywords: river management, assimilatory capacity, dispersion coefficient, sustainable approach, pollution.

1 INTRODUCTION

Preventing pollution of water bodies would result in a better water quality, improved health for humans and aquatic species. However, when this is hampered, their existence will be under threat. With the EU bearing this in mind, a Water Framework Directive (WFD) in 2000 was designed to help chart this course [1], [2]. Even at that, it is herculean to find some of the rivers within the EU member states void of some form of effluent discharge. Therefore, it is important to understand the river dynamics as it relates to mixing and spreading, and this can be achieved by estimating the dispersion coefficient. There are various dispersion mechanism occurring during any form of accidental spill; however, the most predominant is the longitudinal dispersion (LD). According to [3], LD is useful to control biological, chemical and nuclear effluents discharged into rivers. Interestingly, the ID LD models sprung from Taylor in 1954, and more contributions were obtained from Fischer [4], [5]. Thereafter, even more models have evolved. Unfortunately, using these models for prediction do not produce accurate results of LD [7], [8]. Therefore, it is suggested that tracer studies remain one of the reliable methods for the determination of LD in channels, rivers or streams. This method involves the real-time collection of pollutant concentration at the same time or at different times with a section of the river. This pollutant concentration is usually in the form of a conservative tracer – salt tracers, like the one used in this study – and the dispersion coefficient is obtained using different equations and methods and these have been well reported by [9]. However, from the literature search, the various methods implored by various researchers include variable-time constant distance method, variable-time variable distance

WIT Transactions on Ecology and the Environment, Vol 234, © 2019 WIT Press
www.witpress.com, ISSN 1743-3541 (on-line)
doi:10.2495/RBM190201

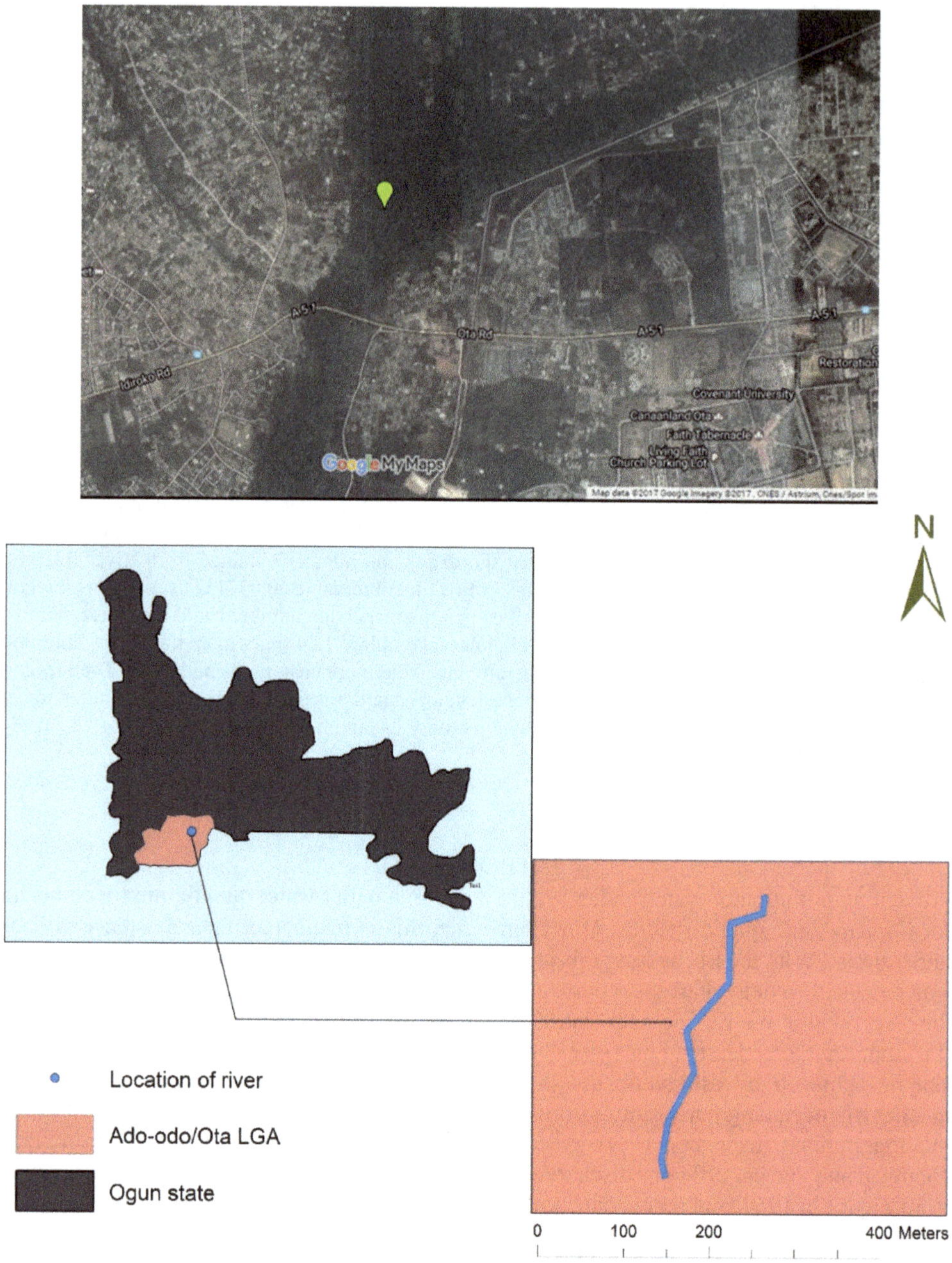

Figure 1: A map showing River Balogun and connecting Rivers within Ota, Ogun State.

method and fixed-time variable distance method [9], [10]. However, so far, the last method aforementioned have been widely used and gives better estimates but does not seem a sustainable approach, with the main challenge of this method is the high cost needed in the purchase of state-of-the-art equipment in large numbers, which is required for tracer data collection. Therefore, this has led to limited studies especially in developing countries on the use of tracers or determination of dispersion coefficients. Agunwamba [10] and Moore [11] reported that there is subjectivity in sampling time interval selection. With no scientific explanation from any researcher on the reasons why the time interval was selected. Therefore, this study aims at the determination of the LD of River Balogun that is under pollution treat from dredging activities and industrial pollution, using the constant distance variable time method, at two different sampling time intervals.

2 METHODS AND MATERIALS

2.1 Study area

The river used for this study is River Balogun with a coordinate of 6°41'6.8"N and 3°8'56.37"E. The river is a tributary from a popular river called River Atuwara. River Balogun is a major source of water for drinking for most villages surrounding Ota, but suffers pollution threat from industries surrounding it. In addition, the inhabitants surrounding the river have major occupation of fishing, bamboo, cutting and dredging. Besides fishing, the other two activities aforementioned can lead to local pollution. Furthermore, in the dry and wet seasons, the volume of water in the river is enough for boating even when packed full of dredged sand.

2.2 Field work

In this study, the constant distance variable time method was employed in the collection of tracer concentration values. 25 kg of locally made salt (Dangote salt) with 97% purity level was used as a conservative tracer at each sampling scheme. It was pre-mixed with the River water in parts, in order increase it solubility when poured into the river. Beforehand, the background concentration was obtained before the River was dosed with the tracer. The reason is to know the actual elevation caused by the salt. A 400 m long portion of the river was marked out for this research using a GPS. After dosing, with the use of a Hanna edge 2020 series Electrical conductivity meter, values were recorded at two different sampling time intervals of 3 minutes and 5 minutes respectively on the same day in the month of January 2017 using the constant distance variable time method otherwise known as the levenspiel and smith approach. This approach involves the monitoring of Electrical conductivity at the outlet at a selected time interval. In this study and in the region where it was conducted, the month of January served as a critical period whereby the stage of the river is low and pollution can be high. In addition, the Electrical conductivity was monitored using a fabricated multi-leveller sampler (Fig. 2). This sampler was designed to obtain the river water samples transversely and vertically simultaneously thereby possibly improving the values of dispersion number, which will be calculated from the average concentration obtained. Thereafter, with the concentration of tracer obtained, the average transverse concentration values were used to determine the dispersion number from eqns (1)–(3) [10]:

$$d=\frac{1}{8}\left(\sqrt{8\delta^2+1}-1\right), \tag{1}$$

where δ^2 is the area under the normalised concentration-time plot, and given as:

$$\delta^2 = \frac{1}{\bar{\theta}^2}\left(\frac{\sum ct^2}{\sum c} - \bar{\theta}^2\right), \quad (2)$$

$$\bar{\theta} = \frac{\sum ct}{\sum c}, \quad (3)$$

$$\bar{\theta} = L/V. \quad (4)$$

Also, where c, t, L, V and $\bar{\theta}$ = tracer concentration, time interval, length of the river, velocity, actual (experimental) and hypothetical detention time of tracer. The detention time of tracer helps to predict the travel time of the tracer, and this can be achieved hypothetically or experimentally. Eqns (3) and (4) give the experimental and hypothetical values of the tracer during tracer studies.

Figure 2: Morphology measurement of one of the sections of River Balogun.

3 RESULT AND DISCUSSION

3.1 Effect of sampling interval of 3 mins on the breakthrough curve and dispersion number

Fig. 4 shows the electrical conductivity readings collected from River Balogun and plotted against time. As earlier said, the time interval used for the collection was 3 mins. In addition, the figure reveals the hydraulic characteristics of the River. From the graph, the lowest value of concentration was observed at 9 mins after tracer injection. Surprisingly, the value recorded at that time was -0.633 µs/cm which resulted from variation in the background

Figure 3: Fabricated multi-leveller sampler for the simultaneous collection of tracer concentrations.

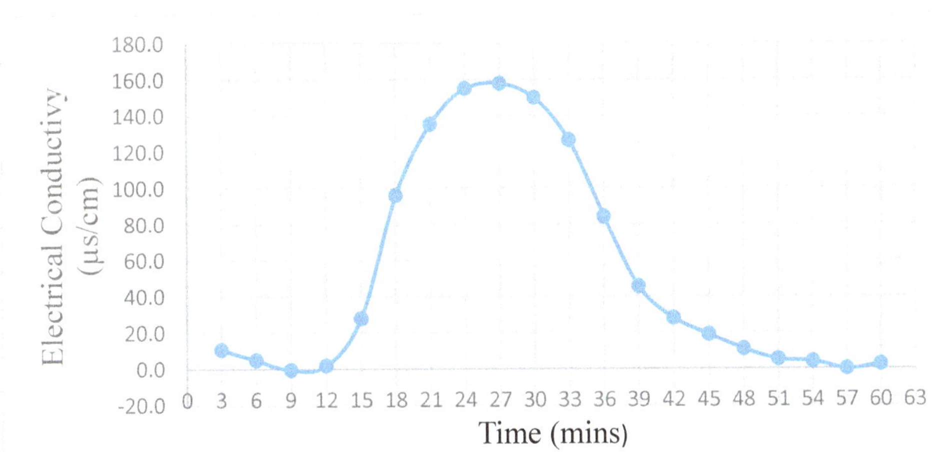

Figure 4: Tracer concentration against time using Levenspiel approach at 3 mins interval.

concentration value with an average value of 48.5 μs/cm which was adopted as the experimental background concentration. However, it was observed that after monitoring the background concentration of the river for some time, some dataset would still produce negative values. This may be due to variation in geochemical properties of the sections of the channel [12], the contribution of other non-point sources that might have increase the background concentration that or due to probe sensitivity. Although, when was noticed, the probe was calibrated more frequently i.e. after each experimental process, but the variation continued. Conversely, it was observed that the tracer was detected close to the theoretical detention time i.e. $L/_V$, from that point onward, consistent increase was observed until it peaked at 27 mins. At 30 mins, a drop in the electrical conductivity was noticed until close to background concentration was obtained at 60 mins of sampling. Furthermore, the graph shows that the long tails usually experienced during sampling is due long sampling time.

3.2 Effect of sampling interval of 5 mins on the breakthrough curve and dispersion number

Fig. 5 also reveals the river pattern in terms of the pollutant movement and shape, which from observation is characterised by the sampling scheme adopted. For instance, the shape of the near Gaussian plot in Figs 4 and 5 are not the same; and would certainly affect the value of dispersion number when calculated. The reason being that the dispersion number is a function of the normalised variance produced by the graphs. Again, in this second sampling scheme, the theoretical detention revealed the point of tracer detection. In addition, unlike the first sampling scheme, there was no negative values observed initially; however, values lower than the difference in the chosen background concentration values were recorded with an average value of 51.3 μs/cm. This is attributed to the reason earlier mentioned for Fig. 4. On the other hand, the peak value in this sampling scheme was observed at 35 mins as against 27 mins in the first sampling scheme. This may be due to increased number of sampling as well as reduced sampling time interval adopted in the first sampling scheme. Specifically, 20 number of tracer readings were collected in the first experimental scheme, while 13 was collected in the second. This is approximately 1.54 times more in the first scheme. However, it has been mathematical established that the more the points available in a graph, the more its accuracy. Therefore, following this thesis, it is assumed that the dispersion number produced by the first scheme should be more accurate than the second scheme. However, this may not be a realistic interval for researchers and some developing countries who still depend manually for data collection and cannot afford the luxury of data loggers. More explicitly, we imply that selecting the same sampling time (3 secs) and any other sampling time lower than this will lead to experimental errors, increase in labour while the intention of reducing sampling time to yield good results would be defeated. Therefore, based on the findings of this study, it is believed that for those researches whereby tracer concentrations were manually or even digitally collected, using any of the known tracer study approaches would have been selected based on the ease of sampling rather than the quest for an accurate value of the dispersion number. Therefore, the result obtained may not be as accurate as expected. Table 1 shows the dispersion number values obtained from the two-different sampling scheme. It reveals that, the dispersion number obtained at 3 mins and that obtained at 5 mins of sampling time interval have a difference of 0.0158 with a factor of 1.74 between it. This shows that sampling time could have a significant effect on the dispersion number generated. However, it is worthy to mention that a variation in the dispersion number can also occur because of the heterogenous nature of river hydrodynamics i.e. Velocity, over time as noticed in this study which varied from $0.12m/s^2$ to $0.46m/s^2$ along and within the studied portion of the river and has been noted as one of the limitations of the constant distance variable time approach as captured in the literature [13]–[15].

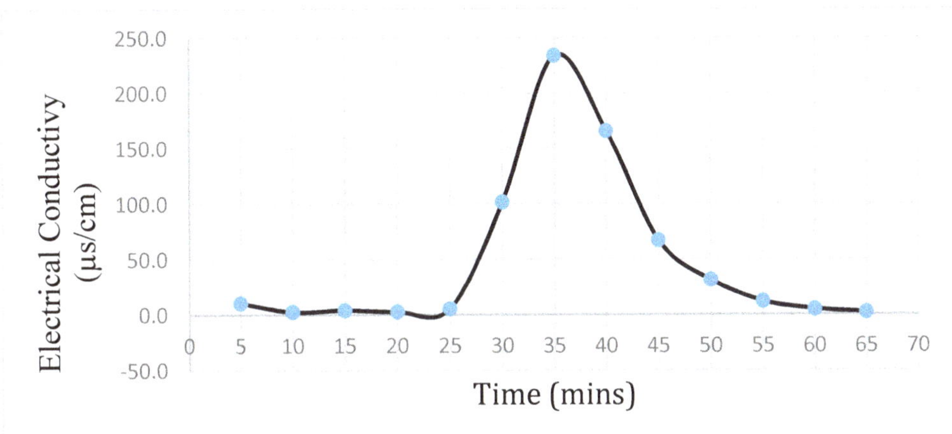

Figure 5: Tracer concentration against time using Levenspiel approach at 5 mins interval.

Table 1: Dispersion number obtained using Levenspiel and Smith Approach at 3 mins and 5 mins respectively.

Dispersion number at 3 mins interval	Dispersion number at 5 mins interval	Difference
0.037152	0.0214	0.0158

4 CONCLUSION

Dispersion number is a useful parameter required in the design of intake structures; determination of intake locations as well as the estimation of assimilatory capacities of Rivers. Therefore, a more realistic value is needed for this purpose, to avoid unnecessary error in design of these structures, avoidance of underestimation or overestimation of this dimensionless parameter. This study employed the Levenspiel and Smith approach to conduct tracer studies for the determination of dispersion number of River Balogun, a tributary of River Atuwara in south-west Nigeria at two different sampling time intervals. Tracer studies were conducted using two different sampling time intervals, and it produced different values for dispersion number with the sampling of 3 mins time interval having a dispersion number 1.76 times greater than the dispersion number produced when the time interval was placed at 5 mins. The reason behind this variance was attributed to variation in sampling time interval as well as the river hydrodynamics. Furthermore, this reveals that the subjectivity in sampling interval practised by different researchers during tracer studies no matter the method has a significant role to play as it could contribute to underestimation or overestimation of the dimensionless parameter. Therefore, it is suggested that, more sustainable approaches should be developed that will require the collection of limited tracer concentration data, reduce sampling time interval errors and consider hydrodynamic variations thereby yielding good estimates of dispersion number which could be better than the Levenspiel and Smith method adopted in this study. This will promote constant monitoring and encourage developing countries to involve in dispersion studies which hitherto have not been frequently practised.

REFERENCES

[1] Velísková, Y., Sokáč, M., Halaj, P., Bara. M.K., Dulovičová, R. & Schügerl, R., Pollutant spreading in a small stream: A case study in Mala Nitra Canal in Slovakia. *Environmental Processes*, **1**, pp. 265–276, 2014.

[2] Velísková, Y., Halaj, P., Sokáč, M. & Bárek, V., Pollution spread analysis in the Malá Nitra River by using of 1-D model. *Acta Horticulturae et Regiotectuare*, **17**, 2014.

[3] Toprak, Z.F., Hamidi, N., Kisi, O. & Gerger, R., Modeling dimensionless longitudinal dispersion coefficient in natural streams using artificial intelligence methods. *KSCE Journal of Civil Engineering*, **18**, pp. 718–730, 2013.

[4] Fischer, B.H., The mechanics of dispersion in natural streams. *J Hydr Div ASCE,* **93**, pp. 187–216, 1967.

[5] Fischer, B.H., Dispersion predictions in natural streams. *J. Sanit. Eng. Div. ASCE*, **94**, pp. 927–943, 1968.

[6] Rankinen, K., Lepistö, A. & Granlund, K., Hydrological application of the INCA model with varying spatial resolution and nitrogen dynamics in a northern river basin. *Hydrology and Earth System Sciences*, pp. 339–350, 2002.

[7] McIntyre, N., Jackson, B., Wade, A., Butterfield, D. & Wheater, H., Sensitivity analysis of a catchment-scale nitrogen model. *Journal of Hydrology*, pp. 71–92, 2005.

[8] Tenebe, I.T., Ogbiye, A.S., Omole, D.O. & Emenike, P.C., Estimation of longitudinal dispersion co-efficient: A review. *Cogent Engineering*, **3**, pp. 216–244, 2016.

[9] Agunwamba, J.C., Reduction of sampling time in tracer studies. *Water Environment Research*, **69**, pp. 343–349, 1997.

[10] Agunwamba, J.C., Optimal design for dispersion experiments. *Water Research*, **2361**, pp. 4570–4582, 2002.

[11] Moore, R., Introduction to salt dilution gauging for streamflow measurement (Part I). *Watershed Management Bulletin*, 2004.

[12] Seo, I.W. & Gadalrah, M.S., Estimating of dispersion coefficient using different forms of lateral velocity distribution. *Proceedings of the WEESHE-99 Conference,* pp. 217–226, 1999.

[13] Perucca, E., Camporeale, C. & Ridolfi, L., Estimation of the dispersion coefficient in rivers with riparian vegetation. *Advances in Water Resources*, **32**, pp. 78–87, 2009.

[14] Shen, C., Niu, J., Anderson, E.J. & Phanikumar, M.S., Estimating longitudinal dispersion in rivers using acoustic doppler current profilers. *Advances in Water Resources*, **33**, pp. 615–623, 2010.

[15] Zeng, Y.H. & Huai, W.X., Estimation of longitudinal dispersion coefficient in rivers. *Journal of Hydro-Environment Research*, **8**, pp. 2–8, 2014.

PRELIMINARY ASSESSMENT OF THE CURRENT POLLUTION STATUS OF THE RIVER ATUWARA, NIGERIA, WITHIN AN INDUSTRIAL SITE: A BIVARIATE APPROACH

OGBIYE A. SAMUEL[1], IMOKHAI T. TENEBE[1,2], PRAISEGOD C. EMENIKE[1], DIWA I. DANIEL[1], DAVID O. OMOLE[1], OMEJE MAXWELL[3], NGENE U. BEN[1], ONYEKWEREDIKE O. KELECHI[4] & IBHADODE OSAGIE[5]
[1]Department of Civil Engineering, Covenant University, Nigeria
[2]Ingram School of Engineering, San Marcos, USA
[3]Department of Physics, Covenant University, Nigeria
[4]Department of Civil Engineering, Federal University, Nigeria
[5]Nigerian Building and Road Research Institute, Nigeria

ABSTRACT

With the increase of industrial growth within developing countries, there is a large tendency for increased pollution within the area. The surrounding rivers may be one of the culprits serving as a waste stream; however, as these rivers are used for agricultural, domestic and recreational purposes, as a cautionary measure, a preliminary assessment of the River Atuwara was carried out to know its status using water quality parameters such as dissolved oxygen (DO), pH, biochemical oxygen demand (BOD), total dissolved solids (TDS), electrical conductivity (EC), coliform count and temperature were examined within certain portions of the River Atuwara, situated in South-West Nigeria. From the study, the descriptive statistics showed that BOD, pH, DO, TDS, EC, coliform count and temperature had mean values of: 3.719±3.34; 6.715±0.48; 4.51±0.49; 113.8±36.67; 174.11±54.62 1036±202 and 25.17±0.313, respectively. The values of pH, DO and coliform count obtained were above the World Health Organization (WHO) and FEPA standards for drinking water. In addition, a principal component analysis (PCA) was conducted on the parameters, which showed that BOD, DO, pH, coliform count and temperature are important parameters to be considered in assessing river water quality. We recommend that adequate treatment of the effluent from the source be carried out as well as using waste treatment plants with natural aerators, to improve the degradation of pollutants. Also, there should be strict follow-up of existing river and watershed management policies, to reduce deliberate and accidental spills.

Keywords: dissolved oxygen, pollution, degradation, river management, dispersion coefficient.

1 INTRODUCTION

Industrialization gained momentum in the 1960s in Africa, as the majority of African countries gained independence: it was a means of self-reliance for many of such brand-new countries, with citizens having the belief that industrialization would pave the way for economic growth and help the continent reach its macroeconomic objectives. This was later achieved to some extent, so much so that human progress these days is expressed in terms of industrial potential and prosperity. Other advantages of industrialization include improvements in the standard of living of the people, availability of goods at cheap prices, availability of options for goods and services, etc.; however, industrialization can bring about unplanned growth, inflation, and displacement of manpower in industries; thereby leading to unemployment, and increased pollution of land, air and water [1], [2].

In recent times, many rivers have fallen victims of pollution that comes from industrial and municipal activities, which causes various imbalances in the river environment and as such, becomes detrimental to the health of the aquatic organisms and humans [3]–[5] who consume the water, unaware of the hazard it constitutes [6]. Conversely, consumers of

WIT Transactions on Ecology and the Environment, Vol 234, © 2019 WIT Press
www.witpress.com, ISSN 1743-3541 (on-line)
doi:10.2495/RBM190211

polluted surface water have the potential to suffer from gastrointestinal, respiratory, dermatological and other forms of disease, which is usually dependent on the concentration of pathogens in the stream, as well as the duration of exposure time [7].

In addition, in a recent study carried out by Emenike et al. [8], it was revealed that most people residing within this river find it less stressful and cost effective to source water from the river. This holds, as there are a lot of problems associated with public boreholes meeting the demands of the public in most rural communities in Nigeria. These problems include an unreliable power supply, which is needed to pump water from the Water Cooperation in that area to other nearby sources were the citizens can access it; a poor distribution network system; and the proximity of the people to these sources, among others. In addition, according to Ezilo and Dune [9] and Adedokun and Agunwamba [10], it is rather unfortunate that most rivers flowing around industrial layouts in Nigeria are usually heavily polluted, due to indiscriminate dumping of wastes by these industries, and the effluents contain pollutants such as heavy metals, organic toxins, solids, and more; however, the pollution of surface water bodies is not limited to industrial effluents. According to several studies [11]–[13], there is a possibility for surface water bodies to become polluted if the environment around them is polluted. Also, the quality and sufficiency of water at the disposal of the people is the measure of their quality of life [14]; therefore, for good water quality, the environment where the river is situated has a role to play. In Ado-Odo/Ota, the local government area where the river is situated and where a study was carried out, there are over 100 large- and small-scale industries, out of which 35 industries discharge untreated or partially treated effluent into the river [14]. Therefore, for a non-toxic and sustainable environment, there is a need for thorough and continuous assessment of the rivers in Nigeria, to ensure the control of waste lies within the assimilative capacity of such rivers [10]. The aim of this study is to conduct a preliminary assessment of the status of the River Atuwara in South-West Nigeria, as it is under threat from industrial pollution and the significance of the pollution must be determined.

2 METHODS AND MATERIALS

2.1 Study area

The River Atuwara, popularly known as the river Iju, passes through the Iju community in Ota, which is an urban industrial center. The Ado-Odo/Ota Local Government Area (LGA) is the most populous LGA in Ogun state. With a population of 526 565 700 330 people, it lies between latitude 60 30'N-60 50'N and longitude 30 02'E-30 25'E [15]. The drainage area of the Atuwara watershed is 147 333 km^2; 50% of which is forested, 10% which is for agriculture, 25% is water, 7% is for residential purposes and 8% for commercial purposes. The River Atuwara has many confluences where other rivers merge with it, as well as tributaries where it feeds into other smaller streams and rivers. The segment of the river under study covers about 1 km, within which various waste discharge points were identified including an industrial effluent discharge point that discharges waste water into the river continuously, all day long.

It was observed that numerous activities are being carried out within the river, such as: the dredging of sand, bathing, laundry; and also there is a water intake structure belonging to the Ogun State water works, which was identified within the watercourse. This intake structure pumps water, which is then treated and distributed to the public taps within the area.

2.2 Field method

This comprises a reconnaissance survey, as well as getting familiar with the river and the environment at large; the determination of the sources of pollution; and the identification of the various uses the river is being put to.

2.3 Sampling visit

Ten sampling stations were marked out for data collection with wooden pegs. The trajectory was marked at intervals of 100 m in a sampling length of 1km. River water samples were collected at those points. Samples were collected before the major effluent inlet point, to determine the effect of the effluent discharge from a popular alcoholic beverage company on the river's water quality. Parameters were determined in-situ; meanwhile, others were determined in the laboratory. Parameters determined in-situ include: pH, temperature, stream velocity, depth, width, dissolved oxygen (DO) and conductivity; while the biochemical oxygen demand (BOD), total dissolved solids (TDS) and coliform count test were carried out in the laboratory, situated in the department of Civil Engineering and the department of biological sciences at Covenant University, respectively. The stream velocity was measured using the global water flow probe, depth was taken along the river at cross-sections using a Speedtech Portable depth sounder, and the water temperature, pH, DO, electrical conductivity (EC) and TDS were measured using the portable Hanna multi-parameter tester (H1 2020 series).

2.4 Laboratory analysis

All laboratory analyses was carried out at Covenant University in Ota Ogun state, Nigeria. The water samples meant for BOD were taken in clean 75 ml plastic bottles and transported immediately to the laboratory where it was incubated at 4°C until the BOD test was carried out. BOD was carried out [16]. Water samples for bacteriological analysis were collected in sterilized containers and kept at a temperature of 4°C; then transported to the biological science laboratory unit in Covenant University for analysis.

2.5 Data analysis

Descriptive statistics and a principal component analysis (PCA) were carried out on all obtained parameters. The PCA used a new set of variables, which it generates as smaller than the original set of variables, but retains the samples' original information [17]. To identify the relationships between various waste water parameters that were present in the river water samples, the data was analyzed using SPSS and a Pearson correlation matrix. Furthermore, Pearson correlation was also used to determine the effect of industrial effluents on the general water quality of the river.

3 RESULTS AND DISCUSSION

The descriptive statistics showed that BOD, pH, DO, TDS, EC, coliform count, temperature and TSS had mean values of: 3.719±3.34 mg/l; 6.715±0.48 mg/l; 4.624±0.49 mg/l; 113.8±36.67 mg/l; 174.11±54.62 µs/lcm, 1036±202 MPN/100 ml and 25.17±0.313°C, respectively; and when compared with the WHO standard for drinking water, it was observed that the mean values of pH, DO and coliform count were above the drinking water standard; hence, the general health of communities along the river is at risk, as it was discovered that the communities around the river drink the water, as well as use it for other domestic purposes without any form of treatment. The sampling points with their corresponding physiochemical properties; as well as some waste water characteristics, are presented in Figs 1–6.

From Fig. 1 below, it is seen that there is a considerable drop in the pH of the water, from 6.69 to 5.53 at point B, suggesting that the studied effluent is acidic, thereby making it unsuitable for both recreation and domestic purposes; furthermore, the acidic nature of the river will affect the conduits' usage for the water supply within the area. The water supply networks would be prone to corrosion of both concrete and metal pipe networks, which would further impair the quality of water supplied within the area. In addition, the increase in pH of the river is responsible for the corresponding high values of EC and TDS that were recorded at point B.

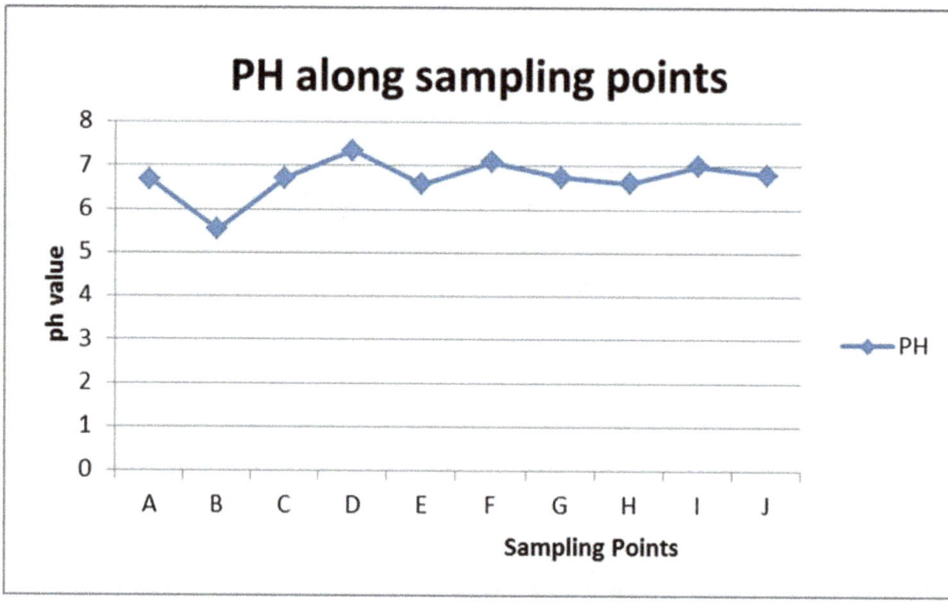

Figure 1: Plot of pH against sampling points.

Figs 2 and 3 show the variability of EC and BOD along the river. They also reveal that EC and BOD were high at point B, as the numbers varied from 159 µs/cm–329 µs/cm and 1.14 mg/l–8.71 mg/l, respectively. The reason for this variation is because the effluent from the alcoholic beverage company situated around the river is discharged at point B, and as the direction of flow is from point A to point B, this shows point A is free from the addition of pollutants by the company; furthermore, this adds up to the reason for the decrease of both parameters, as the flow moves downstream. Fig. 3 also reveals that the river has a strong self-cleaning capacity, as the values of the BOD were reduced greatly along the river, except for point H. The reason could be the presence of other non-point sources of pollution being discharged into the river. In addition, Fig. 4 shows the plot of TDS along the river, and is seen to have a similar relationship with the EC. Conversely, it reveals that the effluent entering the river contains high levels of dissolved solids. Also, the high values of EC and TDS, having values of 329 µs/cm and 218 mg/L, respectively, indicates that the effluent contains a lot of dissolved solids as well as metals [18], [19].

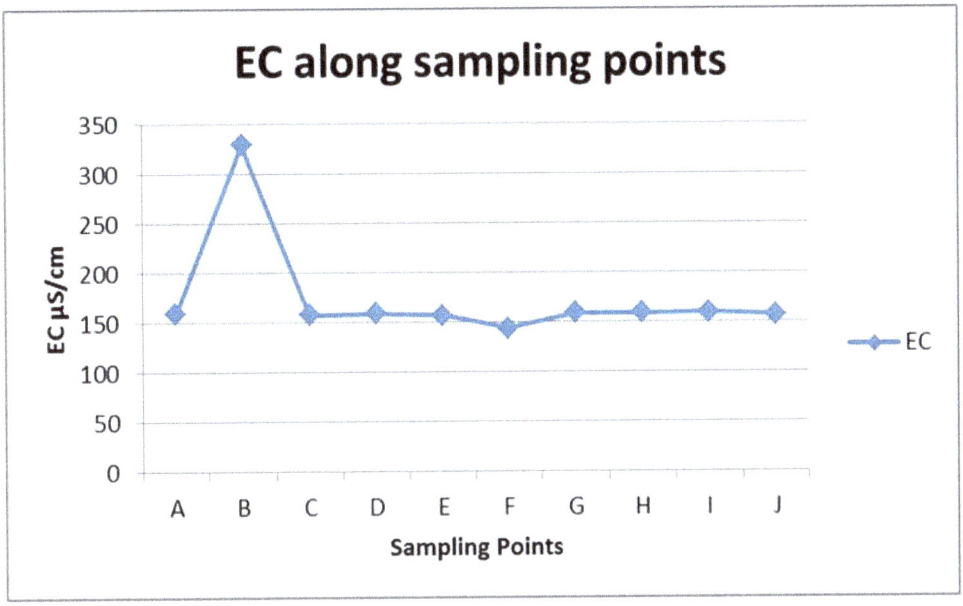

Figure 2: Plot of EC against sample points.

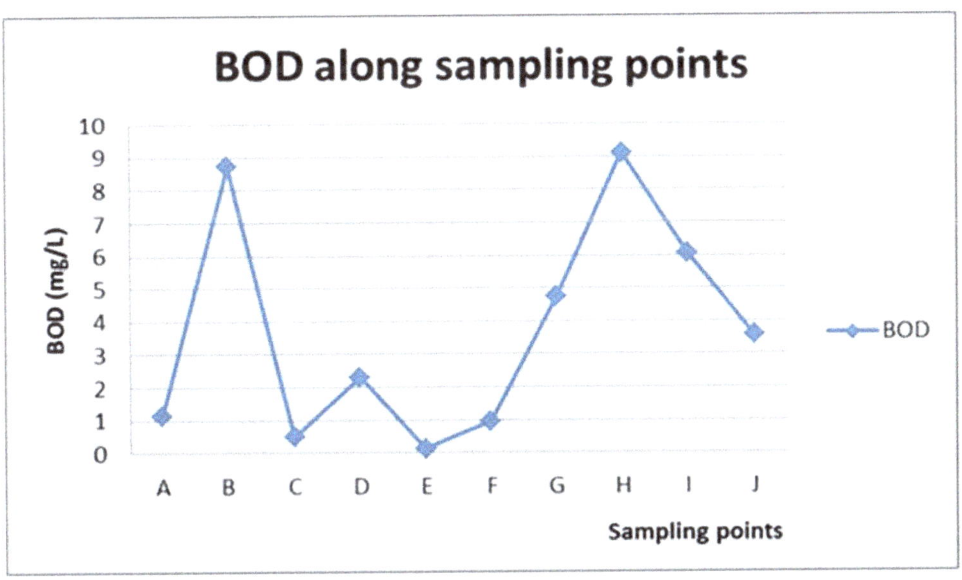

Figure 3: Plot of BOD against sample points.

Fig. 5 presents the plot of DO along the river: it could be observed from this figure that point A and J had the highest and the same DO values of 5.13 mg/L; this shows that all points in the river have DO levels below the expected 7 mg/L. This suggests the level of pollution is on-going in the River Atuwara, with point B having the lowest DO value, 3.97 mg/L. The

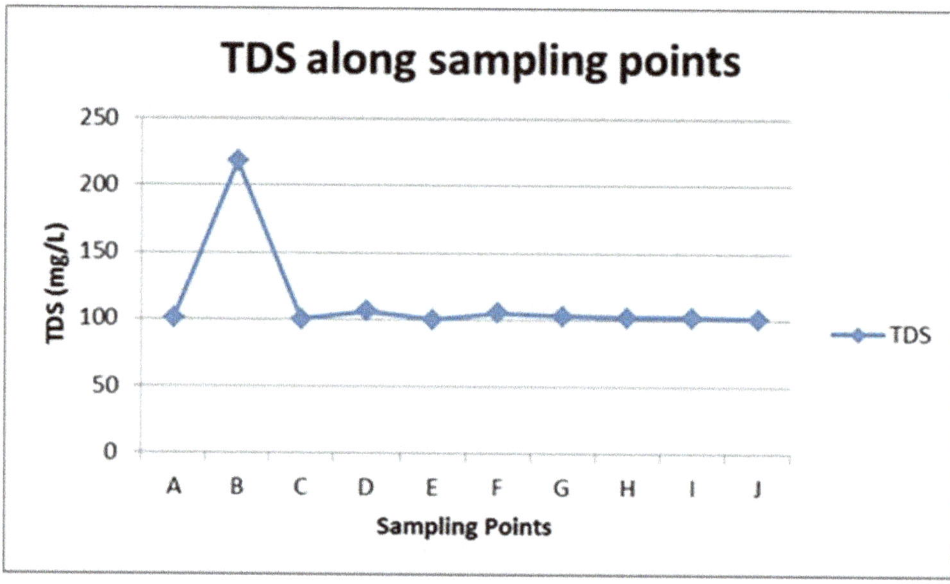

Figure 4: Plot of TDS against sample points.

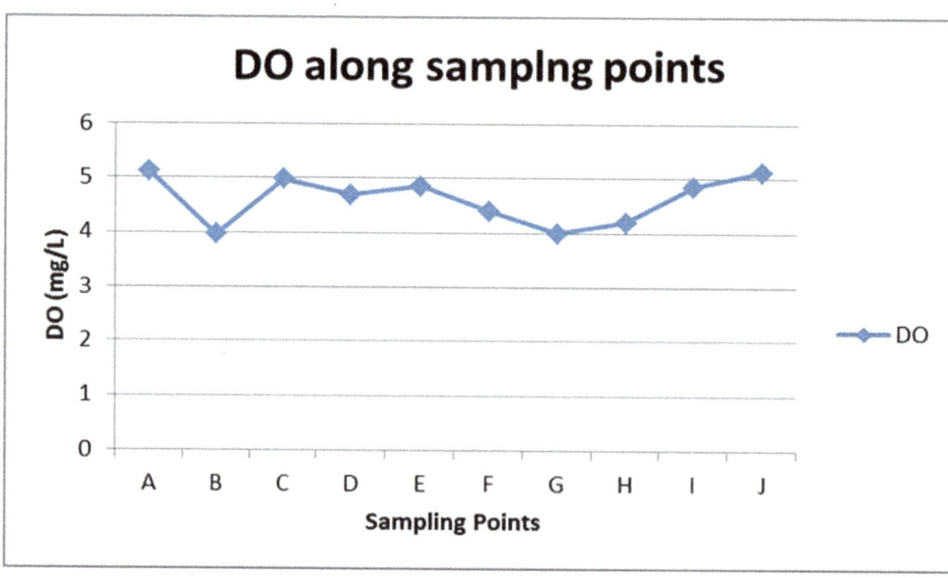

Figure 5: Plot of DO against sample points.

low DO value at point B is recorded because at that point, the alcohol beverage industry within the river vicinity discharges its effluent directly. This change increases the presence of microorganisms; therefore, reducing the available oxygen, as it is utilized for degradation of the effluent. In addition, since the point before the effluent discharged around the study has low DO values, it simply shows that a great portion of the river is under industrial pollution threat. This is because there are several industries on the upstream side of the river that also discharge an uncontrolled volume of waste into the river. That is the reason why even the point before our own study area, Point B, has low DO. Fig. 6 presents the plot of total coliform count along the river; it is observed that just point C has a value of 460 MPN/100 mL, as against the unanimous value of 1100 MPN/100 mL, indicating there is a very high bacteriological waste load in the river. The high amount of microbes present could be attributed to industrial activities; but also human activities such as farming, fishing, bathing, as well as disposal of human and animal wastes into the river.

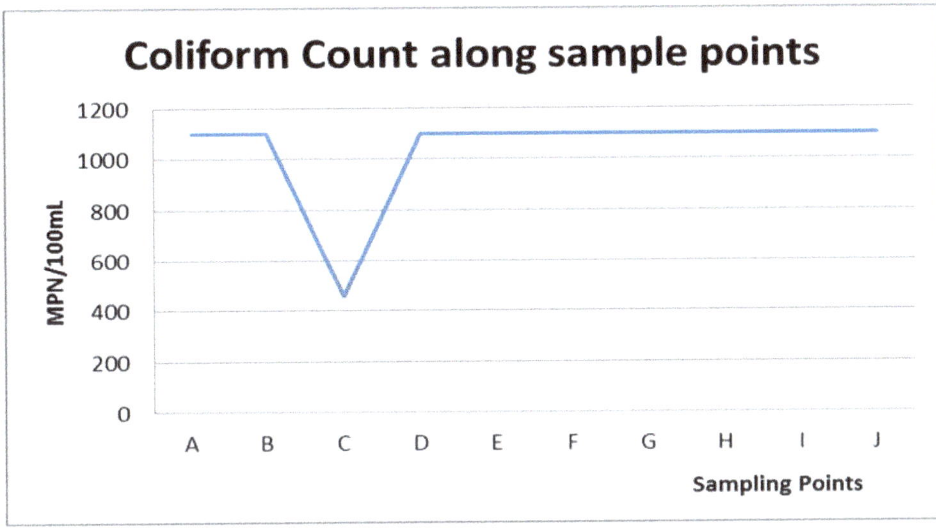

Figure 6: Plot of coliform count against sample points.

PCA was conducted with the set of variables analyzed during the experimental process. This was carried out to identify the contribution of the parameters to the general water quality of the River Atuwara. In Table 1, the variables are grouped into five different groups; however, it is theorized that all of the variables are contributors in the water quality analysis, except temperature. The main function of a PCA is to reduce dimensionality through finding a new set of variables, meanwhile regaining the initial information; hence, in considering

Table 1: Component score coefficient matrix.

	Component				
	1	2	3	4	5
VELOCITY	0.000	-0.003	0.013	1.033	0.660
pH	0.000	-0.004	-0.002	0.013	1.265
TEMP	0.000	0.003	0.010	0.141	-0.276
BOD	0.001	0.326	-1.165	0.275	0.625
TDS	0.309	-7.699	-4.180	-0.754	-0.694
EC	0.692	7.503	4.798	-0.105	1.486
DO	0.000	0.002	0.014	-0.396	0.920

Note: Extraction method: Principal component analysis.

component 5 from Table 1, it was revealed that velocity, pH, BOD, EC, and DO are active contributors in analyzing the water quality of the river with coefficient matrix values: 0.660, 1.265, 0.625, 1.486 and 0.92, respectively. This is because at component 5, all the variables have significant values greater than 0.05.

The Pearson correlation matrix was used to investigate (statistically) the effects of industrial effluents as well as other pollutants on the physicochemical and bacteriological quality of the River Atuwara, by determining the individual contribution and relationships of the variables with each other.

Table 2 below shows the correlation matrix, as well as the relationship between the parameters. From Table 2, it was observed that there is a negative relationship between EC and pH, with a correlation value of -0.877, which implies that an increase in EC would result in a decrease in pH. This relationship is evident from the results obtained in this research. As effluents are discharged into the river, there is a corresponding reduction in the pH, as it becomes more acidic. This pH reduction will threaten aquatic life and reduce water quality. Similarly, pH also has a negative relationship with TDS and from Fig. 4, it is seen that at point B, where the industrial effluent enters the river, the TDS is very high (218 mg/L) and with a corresponding low pH; hence, further justifying the acidity of the effluent. Considering TDS and EC, it is seen that there is a positive correlation of 0.993 between them, this further concludes that the EC value obtained during the water quality check is a function of the dissolved solids found in the water samples.

Table 2: Inter-variable correlation matrix.

	Velocity	pH	Temp	BOD	TDS	EC	DO	Coliform count
Velocity	**1.000**	-0.388	0.174	0.062	0.557	0.547	0.192	-0.417
pH	-0.388	**1.000**	0.053	-0.497	-0.843	-0.877	-0.315	-0.004
Temperature	0.174	0.053	**1.000**	-0.396	-0.335	-0.278	0.206	-0.820
BOD	0.062	-0.497	-0.396	**1.000**	0.529	0.549	-0.184	0.341
TDS	0.557	-0.843	-0.335	0.529	**1.000**	0.993	0.356	0.132
EC	0.547	-0.877	-0.278	0.549	0.993	**1.000**	0.356	0.104
DO	0.192	-0.315	0.206	-0.184	0.356	0.356	**1.000**	-0.338
Coliform count	-0.417	-0.004	-0.820	0.341	0.132	0.104	-0.338	**1.000**

In the same vein, EC and velocity also have a positive correlation between themselves, with a correlation value of 0.547. This implies that high stream velocities would cause the pollutants to travel faster, which is also seen in the relationship between TDS and velocity, with a correlation value of 0.557. Since TDS accounts for the dissolved solids, the correlation simply means that the higher the velocity, the faster these dissolved solids are dispersed in the river. Temperature has a negative correlation of -0.820 with coliform count; hence, an increase of temperature could cause a decrease in the number of coliform bacteria present in the river. In addition, BOD and EC have a positive correlation of 0.549, that means the more pollutants enter the river, the higher the BOD, as well as the conductivity of the river. Similarly, TDS and BOD were seen to have a correlation of 0.529, which implied that an increase in TDS would also cause an increase in BOD.

In order to determine if the observed correlations between the water quality parameters were significant, the data was subjected to paired sample t-tests. The results of a statistical analysis are presented in Table 3. From the test, only correlations present are those having a significant value < 0.05.

Table 3 shows the inter-parameter correlations and their corresponding significance. With the null hypothesis rejected when p > 0.05; it is observed that between EC and pH; pH and TDS; and TDS and EC, the relationships between them are significant. Contrarily, the correlation between EC and velocity; EC and BOD; TDS and velocity and TDS and BOD were all found not to be significant.

Table 3: Paired samples correlations of water quality parameters.

		N	Correlation	Sig.
Pair 1	EC and pH	9	-0.883	0.002
Pair 2	pH and TDS	9	-0.852	0.004
Pair 4	TDS and EC	9	0.993	0.000
Pair 5	EC and Velocity	9	0.541	0.133
Pair 6	EC and BOD	9	0.545	0.129
Pair 7	TDS and Velocity	9	0.548	0.127
Pair 8	TDS and BOD	9	0.519	0.153

4 CONCLUSIONS

We investigated the contribution of industrial effluent on the water quality of the River Atuwara. This was done mainly because for a non-toxic and sustainable environment to exist, there is a need for thorough and continuous assessment of the rivers in Nigeria, to safeguard the control of waste going into the river bodies, within their assimilative capacities. Our results show that the average values of BOD, pH, DO, TDS, EC, coliform count, and temperature were 3.719±3.34 mg/l; 6.715±0.48 mg/l; 4.51±0.49 mg/l; 113.8±36.67 mg/l; 174.11±54.62 µs/cm; 1036±202 MPN/100 ml and 25.17±0.313°C, respectively.

By comparing our values obtained for the Atuwara with the WHO drinking water standard pH, DO and coliform count, they were found to exceed the drinking water standard by 6.5–8.5; 7 mg/L and 10 [20], hence the water was rendered unfit for drinking.

In conclusion a strict, as well as total adherence to the policies already in place is required to improve the water quality status of Nigerian River Atuwara; which is widely used for bathing, drinking, as well as for irrigation purposes. Furthermore, adequate treatment to reduce the organic waste load of the effluent being discharged into the river is recommended. Natural aerators, such as a roughened side, wall and bed enough to cause variation in depth and hydraulic jumps should be integrated into the wastewater treatment plants, to increase the DO content required for the degradation of organisms.

Based on the water quality of the river, stringent laws and enforcement of those laws against indiscriminate disposal of effluent are needed as soon as possible, around the country. This is because the general misconception that the rivers will just take care of the wastes discharged into them does not consider the assimilative capacity of the river, which is a function of the river dynamics, such as velocity, depth, topography, etc. Also, adequate enlightenment should be given to the inhabitants of communities around the river, as to how to use the river water for domestic use. Direct drinking of the water before treatment should be discouraged, as well as indiscriminate dumping of waste by the community inhabitants.

ACKNOWLEDGEMENT

The authors graciously thank the management of Covenant University (in Ota, Ogun State) for continued support of this research.

REFERENCES

[1] Tamuno, S.O. & Edoumiekumo, S.G., Industrialization and trade globalization: What hope for Nigeria? *Int. J. Acad. Res. Bus. Soc. Sci.*, **2**(6), 2012.

[2] Chete, L.N., Adeoti, J.O., Adeyinka, F.M. & Ogundele, O., Industrial development and growth in Nigeria lessons and challenges: Learning to complete, Working Paper No. 8, 2014.

[3] Emenike, P.C., Omole, D.O., Ngene, B.U. & Tenebe, I.T., Potentiality of agricultural adsorbentfor the sequestering of metal ions from wastewater. *Global Journal of Environmental Science and Management*, **2**, pp. 411–442, 2016.

[4] Omole, D.O., Emenike, C.P., Tenebe, I.T., Akinde, A.O. & Badejo, A.A., An assessment of water related diseases in a Nigerian community. *Research Journal of Applied Sciences, Engineering and Technology,* **10**(7), pp. 776–781, 2015.

[5] Tenebe, I.T., Emenike, P.C. & Chukwuka, C.D., Prevalence of heavy metals and computation of its associated risk in surface water consumed in Ado-Odo Ota, South-West Nigeria. *Human and Ecological Risk Assessment: An International Journal,* 2018. DOI: doi.org/10.1080/10807039.2018.1454824.

[6] Tenebe, I.T., Ogbiye, A.S., Omole, D.O. & Emenike, P.C., Estimation of longitudinal dispersion co-efficient: A review. *Cogent Engineering,* **3**, p. 1216244, 2016.

[7] Nzeadibe, T.C., Solid waste reforms and informal recycling in Enugu Urban Area, Nigeria. *Habitat International,* **33**(1), pp. 93–99, 2009.

[8] Emenike, P.C. et al., Accessing safe drinking water in sub-Saharan Africa: Issues and challenges in Southwest Nigeria. *Sustainable Cities and Societies,* **30**, pp. 263–272, 2017.

[9] Ezilo, F. & Dune, K., Effect of environmental pollution on a receiving water body: A case study of Amadi Creek, Port Harcourt, Nigeria. *Transnational Journal of Science and Technology*, **2**(8), 2012.

[10] Adedokun, T.A. & Agunwamba, J.C., Modelling the effect of industrial effluents on water quality: A case study of River Challawa in Nigeria. *International Journal of Modern Engineering Research (IJMER),* **3**, pp. 2797–2803, 2013.

[11] Aliyu, B.N., Municipal solid waste characteristics in three residential zones of Kano Metropolis: Implications for management in the book of proceedings. *50th Annual Conference of the Association of Nigeria Geographers,* University of Calabar, August 25–28, 2008.

[12] Obeta, M.C. & Ochege, F.U., Effects of waste dumps on stream water quality in rural areas of southern Nigeria. *IOSR Journal of Environmental Science, Toxicology and Food Technology (IOSR-JESTFT),* **8**(2), pp. 82–88, 2014.

[13] Ocheri, M.I., Mile, Y. & Oklo, A. Nitrate contamination of shallow wells in Makurdi Urban Area of Benue State. Book of Proceedings. *50th Annual Conference of the Association of Nigeria Geographers,* University of Calabar, August 25–28, 2008.

[14] Adewumi, I.K. & Ogbiye, A.S., Using water hyacinth (*Eichhornia crassipes*) to treat wastewater of a residential university. *Toxicological and Environmental Chemistry,* **91**(5), pp. 891–903, 2009.

[15] Omole, D.O., *Water Quality Modelling: Case Study of the Impact of Abattoir Effluent on River Illo, Ota, Nigeria*, LAP Lambert Academic Publishing: Germany, 2010.

[16] APHA, *Standards Methods for the Examination of Water and Wastewater*, APHA/AWWA/WEF: Washington, DC, 1998.

[17] Camacho, J., Pérez-Villegas, A., Rodríguez-Gómez, R.A. & Jiménez-Mañas, E., Multivariate Exploratory Data Analysis (MEDA) toolbox for Matlab. *Chemometrics and Intelligent Laboratory Systems,* **143**, pp. 49–57, 2015.

[18] Fernandez, B., Seijo, I., Ruiz-Filippi, G., Roca, E., Tarenzi, L. & Lema, J.M., Characterization management and treatment of wastewater from white wine production. *Water Science and Technology,* **56**(2), pp. 121–128, 2007.

[19] Omole, D.O., Adewumi, I.K., Longe, E.O. & Ogbiye, A.S., Study of auto purification capacity of River Atuwara. *Nigeria International Journal of Engineering and Technology*, **2**(2), pp. 229–235, 2012.

[20] World Health Organization (WHO), *Guidelines for Drinking Water Quality*, 4th ed., WHO: Geneva, Switzerland, 2011.

Author index

www.ingramcontent.com/pod-product-compliance
Lightning Source LLC
LaVergne TN
LVHW062236200926
848484LV00006B/1551
9781784663278